Android Wear Projects

Create smart Android Apps for Wearables

Ashok Kumar S

BIRMINGHAM - MUMBAI

Android Wear Projects

First published: July 2017

Production reference: 1260717

Published by Packt Publishing Ltd.
Livery Place
35 Livery Street
Birmingham
B3 2PB, UK.
ISBN 978-1-78712-322-9

www.packtpub.com

Credits

Author
Ashok Kumar S

Reviewers
Ravindra Kumar
Natarajan Raman

Commissioning Editor
Amarabha Banerjee

Acquisition Editor
Nigel Fernandes

Content Development Editor
Mohammed Yusuf Imaratwale

Technical Editor
Rashil Shah

Copy Editor
Akshata Lobo

Project Coordinator
Ritika Manoj

Proofreader
Safis Editing

Indexer
Mariammal Chettiyar

Graphics
Jason Monteiro

Production Coordinator
Shantanu Zagade

About the Author

Ashok Kumar S is an Android developer residing in Bangalore, India. A gadget enthusiast, he thrives on innovating and exploring the latest gizmos in the tech space.

He has been developing Android applications for all Google-related technologies. He is a Google certified Udacity Nano degree holder.

A strong believer in spirituality, he heavily contributes to the open source community as a e-spiritual ritual to improve his e-karma. He regularly conducts workshops about Android application development in local engineering colleges and schools. He also organizes multiple tech events at his organization and is a regular attendee of all the conferences that happen on Android in the silicon valley of India.

He also has a YouTube channel, called *AndroidABCD*, where he discusses all aspects of Android development, including quick tutorials.

I would like to thank Mr. Mohan B A for guiding me in everything I plan to accomplish; he's been my strength and a great teacher.

I would like to thank my colleagues at Dunst Technologies Pvt. Ltd for supporting me and for encouraging me to write this book. Dunst has always been a great place to learn and implement everything I learned in the real world. It is a privilege to be a part of an outstanding organisation such as Dunst.

I would like to thank my family for all their support, especially my mother Lalitha, sister-in-law Sumithra, and Krishna for constantly pushing me to do my best and making sure that I would never go hungry to bed.

I would like to thank all the people who have supported me at every stage of this book, especially Vinod S Nair and Ashwin R Nair.

I would like to thank the Packt Publishing team for this opportunity and for supporting me throughout my journey to complete this book.

About the Reviewers

Ravindra Kumar is an Android developer and computer polyglot from Bengaluru, India. He is an Android and web speaker, startup geek, and open source junkie.

He works as an Android developer at Fueled. Previously, he worked with Cleartrip, as the lead developer of Cleartrip.com's Android app. He likes open source projects and is a huge fan of the fancy Android libraries out there. He contributes to bug reporting, fixing, and feedback, and has given talks at DroidCon, TiConf, and JSFOO.

He started as a web engineer who used to write lots of JavaScript, but after some time, looking for his real passion, he started his journey in mobile app development through Titanium. Later, he discovered the Android world. After getting some experience on such an awesome platform, he started a new adventure at a mobile company, where he led several projects for important Indian companies.

He has a strong interest in code quality, testing, and automation, and preferably all three together. Combining technical with soft skills, he also ventures into the realms of mentoring and article writing. He hates doing things manually, and hates to see src/test/java directories with empty example classes. He believes that by working with legacy code and improving it, he can make the world a better place. To his disappointment, the world does not seem to care all that much about his efforts.

He is a pretty normal person--a husband, father of one, and lover of cricket. You can follow `@ravidsrk` on Twitter or email him at `ravidsrk@gmail.com`.

> *I'm thankful to my wife, Abhilasha, and my son, Hemanth, for understanding that I couldn't be there at times while I was helping review this book.*

Natarajan Raman has over 13 years of experience in software design and development with a good understanding and experience of the complete SDLC life cycle. Natarajan Raman is a Google/Udacity certified Nano degree holder on Android development. He was invited as a guest by Google to the I/O 2017, which was held in May in San Francisco, California.

Natarajan works for the Patterns design school and is also the managing trustee of Dream India, `dreamindia.org`, an NGO started by youngsters inspired by Dr. Kalam.

> *I would like to thank my employer, my colleagues, my family, and the children of Vasantham--`www.vasantham.org`--for being a source of inspiration and support forever.*

www.PacktPub.com

For support files and downloads related to your book, please visit www.PacktPub.com.

Did you know that Packt offers eBook versions of every book published, with PDF and ePub files available? You can upgrade to the eBook version at www.PacktPub.com and as a print book customer, you are entitled to a discount on the eBook copy. Get in touch with us at service@packtpub.com for more details.

At www.PacktPub.com, you can also read a collection of free technical articles, sign up for a range of free newsletters and receive exclusive discounts and offers on Packt books and eBooks.

https://www.packtpub.com/mapt

Get the most in-demand software skills with Mapt. Mapt gives you full access to all Packt books and video courses, as well as industry-leading tools to help you plan your personal development and advance your career.

Why subscribe?

- Fully searchable across every book published by Packt
- Copy and paste, print, and bookmark content
- On demand and accessible via a web browser

Customer Feedback

Thanks for purchasing this Packt book. At Packt, quality is at the heart of our editorial process. To help us improve, please leave us an honest review on this book's Amazon page at `https://www.amazon.com/dp/1787123227`. If you'd like to join our team of regular reviewers, you can e-mail us at `customerreviews@packtpub.com`. We award our regular reviewers with free eBooks and videos in exchange for their valuable feedback. Help us be relentless in improving our products!

Table of Contents

Preface

Android Wear 2.0 is a powerful platform for wearable smart devices. Wear 2.0 enabled close to 72 percent new device activation in the wearable device market from the day Wear was announced. Google is working with multiple iconic brands to bring the best user experience to smart watches. Continuous improvement in the Wear hardware for new devices in the market shows the potential Wear devices can offer. Google introduced a whole new way of experiencing wearable technology with material design, standalone applications, watch face innovations, and more.

The Wear platform is becoming more popular, and Android developers can reap the benefits of improving their ability to program for Wear devices.

This book helps to create five Wear applications, with comprehensive explanations. We start with making a wearable note taking application by exploring Wear-specific user interface components and building a Wear map application with the ability to persist a quick note on the map. We will build a complete chat application with a companion mobile application. We will also build a health and fitness application to monitor pulse rate, a reminder to drink water, and so on, and also write a digital watch. We will complete the book by exploring the capabilities of the Wear 2.0 platform.

Have fun building great wear applications.

What this book covers

Chapter 1, *Getting You Ready to Fly - Setting Up Your Development Environment*, teaches you to write your first Wear application, explores the essential UI components specific to Wear applications, and discusses Android Wear design principles.

Chapter 2, *Let us Help Capture What is on Your Mind - WearRecyclerView and More*, covers `WearableRecyclerView` and the `WearableRecyclerView` adapter, and `SharedPreferences`, `BoxInsetLayout`, and the animated `DelayedConfirmation`.

Chapter 3, *Let us Help Capture What is on Your Mind - Saving Data and Customizing the UI*, explores the integration of the Realm database and custom fonts, UI updates, and finalizing the project.

Chapter 4, *Measure Your Wellness - Sensors*, showcases the accuracy of the sensors, battery consumption, Wear 2.0 doze mode, material design, and so on.

Chapter 5, *Measure Your Wellness - Syncing Collected Sensor Data*, focuses on syncing collected sensor data, collecting sensor data from a Wear device, processing the received data to find calories and distance, sending data to a Wear application from a mobile application, Realm DB integration, `WearableRecyclerView`, and `CardView`.

Chapter 6, *Ways to Get Around Anywhere - WearMap and GoogleAPIclient*, explains the Developer API console; the Maps API Key; and SHA1 Fingerprint, SQlite integration, Google Maps, the Google API Client, and Geocoder.

Chapter 7, *Ways to Get Around Anywhere - UI controls and More*, looks at understanding UI controls, marker controls, map zoom controls, StreetView in Wear, and best practices.

Chapter 8, *Let us Chat in a Smart Way - Messaging API and More*, discusses configuring Firebase for your mobile application, creating a user interface, understanding the messaging API, working with Google API Client, and building a Wear module.

Chapter 9, *Let us Chat in a Smart Way - Notifications and More*, covers Firebase functions, notifications, material design Wear app Wear 2.0 input method framework, and so on.

Chapter 10, *Just a Face for Your Time - WatchFace and Services*, outlines `CanvasWatchFaceService` and registering a watch face, `CanvasWatchFaceService.Engine` and callbacks, watch face elements and initializing them writing the watch face, and handling gestures and tap events.

Chapter 11, *More About Wear 2.0*, explores a standalone application, curved layout and more UI components the Complications API, different navigations and actions, wrist gestures, input method framework, and distributing Wear apps to the Play Store.

What you need for this book

To be able to follow with this book, you need a computer with the latest Android Studio version installed. You need internet to set up all the required SDK for Wear development. If you have a Wear device to test the application, that would be good; otherwise, Android Wear emulators will do the magic.

Who this book is for

This book is for Android developers who already have a strong understanding of programming and developing apps in Android. This book helps the reader advance from being an intermediate-level to an expert-level Android developer, by adding Wear development skills to their knowledge.

Conventions

In this book, you will find a number of text styles that distinguish between different kinds of information. Here are some examples of these styles and an explanation of their meaning.

Specific commands or tools from the interface will be identified as follows:

Select the **Save** button.

Code words in text, database table names, folder names, filenames, file extensions, pathnames, dummy URLs, user input, and Twitter handles are shown as follows: "We can include other contexts through the use of the `include` directive."

A block of code is set as follows:

```
compile 'com.google.android.support:wearable:2.0.0' compile
'com.google.android.gms:play-services-wearable:10.0.1' provided
'com.google.android.wearable:wearable:2.0.0'
```

When we wish to draw your attention to a particular part of a code block, the relevant lines or items are set in bold:

```
<?xml version="1.0" encoding="utf-8"?>
<android.support.wearable.view.BoxInsetLayout
xmlns:android="http://schemas.android.com/apk/res/android"
xmlns:app="http://schemas.android.com/apk/res-auto"
xmlns:tools="http://schemas.android.com/tools"  android:id="@+id/container"
android:layout_width="match_parent"  android:layout_height="match_parent"
tools:context="com.ashok.packt.wear_note_1.MainActivity"
tools:deviceIds="wear"> </android.support.wearable.view.BoxInsetLayout>
```

Any command-line input or output is written as follows:

```
adb connect 192.168.1.100
```

New terms and important words are shown in bold. Words that you see on the screen, for example, in menus or dialog boxes, appear in the text like this: "Let the default selected template be the Wear application code stub **Always On Wear Activity.**"

Warnings or important notes appear like this.

Tips and tricks appear like this.

Reader feedback

Feedback from our readers is always welcome. Let us know what you think about this book-what you liked or disliked. Reader feedback is important for us as it helps us develop titles that you will really get the most out of. To send us general feedback, simply e-mail feedback@packtpub.com, and mention the book's title in the subject of your message. If there is a topic that you have expertise in and you are interested in either writing or contributing to a book, see our author guide at www.packtpub.com/authors.

Customer support

Now that you are the proud owner of a Packt book, we have a number of things to help you to get the most from your purchase.

Downloading the example code

You can download the example code files for this book from your account at http://www.packtpub.com. If you purchased this book elsewhere, you can visit http://www.packtpub.com/support and register to have the files e-mailed directly to you. You can download the code files by following these steps:

1. Log in or register to our website using your e-mail address and password.
2. Hover the mouse pointer on the **SUPPORT** tab at the top.
3. Click on **Code Downloads & Errata**.
4. Enter the name of the book in the **Search** box.
5. Select the book for which you're looking to download the code files.
6. Choose from the drop-down menu where you purchased this book from.
7. Click on **Code Download**.

Once the file is downloaded, please make sure that you unzip or extract the folder using the latest version of:

- WinRAR / 7-Zip for Windows
- Zipeg / iZip / UnRarX for Mac
- 7-Zip / PeaZip for Linux

The code bundle for the book is also hosted on GitHub at `https://github.com/PacktPublishing/Android-Wear-Projects`. We also have other code bundles from our rich catalog of books and videos available at `https://github.com/PacktPublishing/`. Check them out!

Errata

Although we have taken every care to ensure the accuracy of our content, mistakes do happen. If you find a mistake in one of our books-maybe a mistake in the text or the code-we would be grateful if you could report this to us. By doing so, you can save other readers from frustration and help us improve subsequent versions of this book. If you find any errata, please report them by visiting `http://www.packtpub.com/submit-errata`, selecting your book, clicking on the **Errata Submission Form** link, and entering the details of your errata. Once your errata are verified, your submission will be accepted and the errata will be uploaded to our website or added to any list of existing errata under the Errata section of that title. To view the previously submitted errata, go to `https://www.packtpub.com/books/content/support`and enter the name of the book in the search field. The required information will appear under the **Errata** section.

Piracy

Piracy of copyrighted material on the internet is an ongoing problem across all media. At Packt, we take the protection of our copyright and licenses very seriously. If you come across any illegal copies of our works in any form on the Internet, please provide us with the location address or website name immediately so that we can pursue a remedy. Please contact us at `copyright@packtpub.com` with a link to the suspected pirated material. We appreciate your help in protecting our authors and our ability to bring you valuable content.

Questions

If you have a problem with any aspect of this book, you can contact us at `questions@packtpub.com`, and we will do our best to address the problem.

1
Getting You Ready to Fly - Setting Up Your Development Environment

The culture of Wearing a utility that helps us to perform certain actions has always been part of a modern civilization. Wrist watches for human beings have become an augmented tool for checking the time and date. Wearing a watch lets you check the time with just a glance. Technology has taken this watch-wearing experience to the next level. The first modern Wearable watch was a combination of a calculator and a watch, introduced to the world in 1970. Over the decades, advancements in microprocessors and wireless technology have led to the introduction of a concept called *ubiquitous computing*. During this time, most leading electronics industry start-ups started to work on their ideas, which has made Wearable devices very popular.

Tech giant companies, such as Google, Apple, Samsung, and Sony, have joined the force of the Wearable devices era. They have introduced their competitive Wearable products, which are extremely successful in the Wearable device market. More interestingly, Google's Android Wear is powerful, follows the same Android smartphone development practices, and has a very good developer community compared to Apple Watch OS and Samsung's Tizen OS developer community.

Google announced Android Wear in March 2014. Since then, Android Wear as a smartwatch and Wearable software platform has evolved. Google's continuous advancement in designing and user experience have resulted in a new generation of the Android Wear operating system, which has the ability to handle biometric sensors like never before with more features in the platform; Google calls it Android Wear 2.0.

Android Wear 2.0 will cause a lot of excitement in app development with remarkably competitive features to develop. Android Wear 2.0 allows a developer to build and carve his idea specific to Android Wear; there is no need to pair a watch and mobile app. Google calls it a standalone application. Android Wear 2.0 introduces a new way to input within the Android watch: a new application programming interface called Complications, which allows watch faces to display vital information from biometrics and other sensors. New updated notifications support for Android Wear 2.0 will help users and developers to present notifications in a more comprehensive manner.

In this chapter, we will explore the following:

- Android Wear design principles
- Exploring essential UI components specific to Wear apps
- Setting up a development environment for Wear apps development
- Creating your first Android Wear application

Android Wear design principles

Designing a Wear application is different than designing a mobile or tablet application. The Wear operating system is very lightweight and has a specific set of jobs to accomplish by sharing the right information with the Wearer.

General Wear principles are Timely, Glanceable, Easy to Tap, Time-Saving.

Timely

Giving the right information at the right time.

Glanceable

Keeping the Wear application user interface clean and uncluttered.

Easy to Tap

The actions users will click on should have the right spacing and size of the picture.

Time-Saving

Creating the best application flows that do tasks quickly.

For any Wear application, we need the proper building blocks to control the business logic of the application and other architectural implementation. The following are the scenarios for developing a Wear application to help us to carve the wear application better:

- Defining layouts
- Creating lists
- Showing confirmations
- Wear navigation and actions
- Multifunction buttons

Defining layouts

Wearable applications can use the same layouts that we use in handheld Android device programming but with specific constraints for Wear applications. We should not do heavy processing actions similar to handheld Android devices in Wear applications and expect a good user experience.

An application designed for a round screen will not look great on square Wear devices. To resolve this, the Android Wear support library comes with the following two solutions:

- `BoxInsetLayout`
- `Curved Layout`

We can provide different resources to allow Android to detect the shape of the Android Wear at runtime.

Creating lists

Lists let the user select an item from a set of items. In the legacy Wear, 1.x API `WearableListView` helped programmers to build lists and custom lists. Wearable UI library now has `WearableRecyclerView` with `curvedLayout` support and has the best implementation experience in Wear devices.

We can add gestures and other magnificent functionalities:

Exploring UI components for Wear devices

In this subchapter, let's explore the commonly used Wear-specific UI components. In Wear application programming, we can use all the components that we use in mobile app programming, but how we accommodate the visual appearance of components in the Wear device needs to be well thought of before using it.

WatchViewStub: WatchViewStub helps in rendering the views for different form factors of Wearable devices. If your application is being installed on a round watch device, WatchViewStub will load the specific layout configuration for round watches. If it is square, it will load the square layout configuration:

```xml
<?xml version="1.0" encoding="utf-8"?>
<android.support.wearable.view.WatchViewStub
xmlns:android="http://schemas.android.com/apk/res/android"
    xmlns:app="http://schemas.android.com/apk/res-auto"
    xmlns:tools="http://schemas.android.com/tools"
    android:id="@+id/watch_view_stub"
    android:layout_width="match_parent"
    android:layout_height="match_parent"
    app:rectLayout="@layout/rect_activity_main"
    app:roundLayout="@layout/round_activity_main"
    tools:context="com.ashokslsk.wearapp.MainActivity"
    tools:deviceIds="wear"></android.support.wearable.view.WatchViewStub>
```

WearableRecyclerView: WearableRecyclerView is the implementation of recyclerview specific to wearable devices. It provides a flexible view for datasets in the Wearable device viewport. We will explore WearbaleRecyclerView in detail in the coming chapters:

```xml
<android.support.wearable.view.WearableRecyclerView
    android:id="@+id/recycler_launcher_view"
    android:layout_width="match_parent"
    android:layout_height="match_parent"
    android:scrollbars="vertical" />
```

Note: `WearableListView` is deprecated; the Android community recommends using `WearableRecyclerView`.

`CircledImageVIew`: An `Imageview` surrounded by a circle. A very handy component for presenting the image in round form factor Wearable devices:

```xml
<?xml version="1.0" encoding="utf-8"?>
<android.support.wearable.view.CircledImageView
    xmlns:android="http://schemas.android.com/apk/res/android"
    xmlns:app="http://schemas.android.com/apk/res-auto"
    android:id="@+id/circledimageview"
    app:circle_color="#2878ff"
    app:circle_radius="50dp"
    app:circle_radius_pressed="50dp"
    app:circle_border_width="5dip"
    app:circle_border_color="#26ce61"
    android:layout_marginTop="15dp"
    android:src="@drawable/skholinguaicon"
    android:layout_gravity="center_horizontal"
    android:layout_width="wrap_content"
    android:layout_height="wrap_content" />
```

`BoxInsetLayout`: This Layout extends directly to `Framelayout` and it has the ability to recognize the form factor of the Wearable device. Shape-aware `FrameLayout` can box its children in the center square of the screen:

```xml
<android.support.wearable.view.BoxInsetLayout
    xmlns:android="http://schemas.android.com/apk/res/android"
    xmlns:tools="http://schemas.android.com/tools"
    xmlns:app="http://schemas.android.com/apk/res-auto"
    android:layout_width="match_parent"
    android:layout_height="match_parent"
    tools:context="com.example.ranjan.androidwearuicomponents.BoxInsetLayoutDem
o">

    <TextView
        android:text="@string/hello_world"
        android:layout_width="wrap_content"
        android:layout_height="wrap_content"
```

```
            app:layout_box="all" />
```

```
    </android.support.wearable.view.BoxInsetLayout>
```

After the Wear 2.0 release, a few components were deprecated for an immersive activity experience and Google strictly prohibits using them; we can still use all the components that we know in Android programming.

Showing confirmations

Compared to confirmations in handheld Android devices, in Wear applications, confirmations should occupy the whole screen or more than what handheld devices show as a dialogue box. This ensures users can see these confirmations at one glance. The Wearable UI library helps in displaying confirmation timers and animated timers in Android Wear.

DelayedConfirmationView

A `DelayedConfirmationView` is an automatic confirmation view based on the timer:

```
<android.support.wearable.view.DelayedConfirmationView
    android:id="@+id/delayed_confirm"
    android:layout_width="40dp"
    android:layout_height="40dp"
    android:src="@drawable/cancel_circle"
    app:circle_border_color="@color/lightblue"
    app:circle_border_width="4dp"
    app:circle_radius="16dp">
</android.support.wearable.view.DelayedConfirmationView>
```

Wear navigation and actions

In the new release of Android Wear, the **Material design** library adds the following two interactive drawers:

- Navigation drawer
- Action drawer

Navigation drawer

Lets user switch between views in the application. Developers can allow the drawer to be opened anywhere within the scrolling parent's content by setting the `setShouldOnlyOpenWhenAtTop()` method to false:

```
<android.support.wearable.view.drawer.WearableNavigationDrawer
    android:id="@+id/top_drawer"
    android:layout_width="match_parent"
    android:layout_height="match_parent"
    android:background="@android:color/holo_red_light"
    app:navigation_style="single_page"/>
```

Action drawer

The action drawer gives access to easy and common actions in your application. By default, action drawer appears at the bottom of the screen and provides specific actions to users:

```
<android.support.wearable.view.drawer.WearableActionDrawer
    android:id="@+id/bottom_drawer"
    android:layout_width="match_parent"
    android:layout_height="match_parent"
    android:background="@android:color/holo_blue_dark"
    app:show_overflow_in_peek="true"/>
```

Multifunction buttons

In addition to the power button, Android Wear supports another button called the multifunction button on the device. The Wearable support library provides API for determining the multifunction buttons included by the manufacturer:

```
@Override
// Activity
public boolean onKeyDown(int keyCode, KeyEvent event){
   if (event.getRepeatCount() == 0) {
     if (keyCode == KeyEvent.KEYCODE_STEM_1) {
       // Do stuff
       return true;
     } else if (keyCode == KeyEvent.KEYCODE_STEM_2) {
       // Do stuff
       return true;
     } else if (keyCode == KeyEvent.KEYCODE_STEM_3) {
       // Do stuff
       return true;
     }
   }
   return super.onKeyDown(keyCode, event);
}
```

Visit https://developer.android.com/training/wearables/ui/index.html for any sort of query that you might have on design guidelines for Wear device programming.

Setting up a development environment for Wear development

In this section, we will set up a development environment for Wear application development.

Prerequisites

1. Your favorite operating system (Windows, macOS, or Linux)
2. Determine whether you have the latest JRE installed on your operating system
3. Install the latest version of JDK or Open JDK
4. Install the latest version of Android Studio (at the time of writing this book, the latest version is 2.2.3 and any newer version should be fine)

Installing Android Studio

Visit `https://developer.android.com/studio/index.html` to download the latest version of Android Studio. Google highly recommends using Android Studio for all Android application development, since Android Studio has tight integration with Gradle and useful Android APIs:

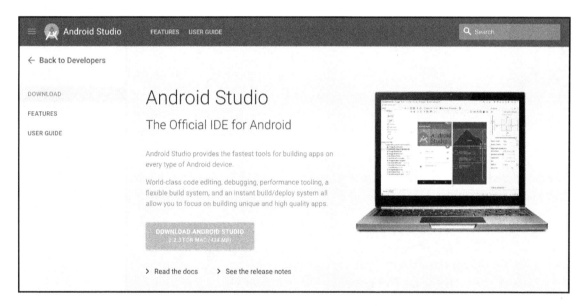

After the Android Studio installation, it's now time to download the necessary SDK in the **SDK Platforms** tab in **SDK Manager**. Install one complete version of Android; for the scope of this book, we will install **Android 7.1.1 API level 25**:

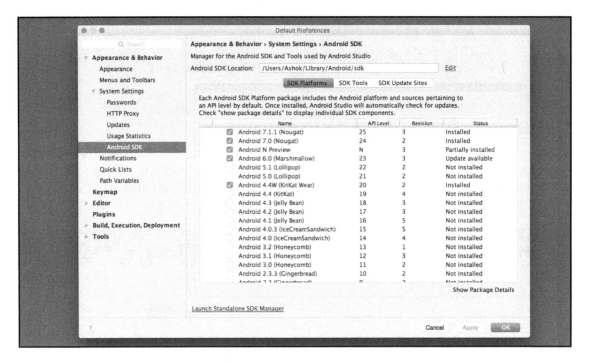

After the successful installation of the SDK of Nougat 7.1.1 API level 25, under the **SDK Tools** tab, make sure you have installed the following components, as shown in the following screenshot:

- **Android Support Library**
- **Google Play services**
- **Google Repository**
- **Android Support Repository**

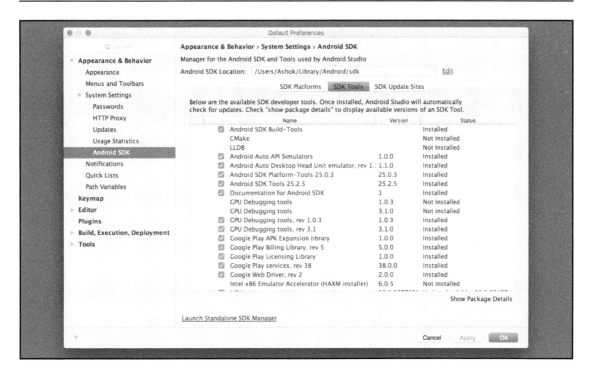

Google releases updates on IDE and SDK Tools frequently; keep your development environment up-to-date.

Note: if you plan to make your application available in China, then you must use the special release version 7.8.87 of the Google Play services client library to handle communication between a phone and watch: `https://developer.android.com/training/wearables/apps/creating-app-china.html`

Visit the following link to check the update **Release Notes** on **SDK Tools**: `https://develo per.android.com/studio/releases/sdk-tools.html`.

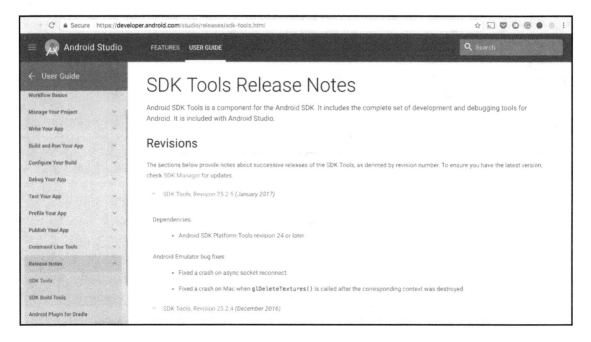

Updating your IDE from the stable channel is highly recommended. Updates for Android Studio are available on four different channels:

- Canary channel
- Dev channel
- Beta channel
- Stable channel

Canary channel: The Android Studio engineering team works continuously to make Android Studio better. In this channel, every week there will be an update release, and it will include new functionality changes and improvements; you can check those changes in the release notes. But updates from this channel are not recommended for application production.

Dev Channel: On this channel, a release happens after a complete round of internal testing from the Android Studio team.

Beta channel: On this channel, updates are totally based on stable Canary builds. Before publishing these builds to a stable channel, Google releases them in the beta channel to get developer feedback.

Stable Channel: Are the official stable releases of the Android Studio and will be available to download on Google's official page `http://developer.android.com/studio.`

By default, Android Studio receives updates from a stable channel.

Creating your first Android Wear application

In this section, let's understand the essential steps required to create your first Wear project.

Before you continue to create your application, ensure you have one complete version of Android installed with a Wear system image and you have the latest version of Android Studio.

The following picture is the initial interface of Android Studio. In this window, one can import legacy ADT Android projects, configure the Android SDK, and update Android Studio.

Android Studio welcome window with basic controls for getting started:

Creating your first Wear project

Click on the **Start a new Android Studio project** option in the Android Studio window. You will be prompted by another window with project details.

The following screenshot shows the window that allows users to configure their project details, such as project name, **Package name**, and whether the project needs native C++ support:

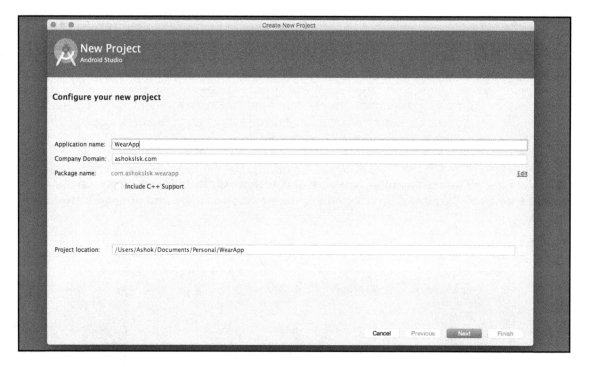

You can name your project as you wish. After you have chosen your project name and your project local system location, you can press the **Next** button in the window, which brings up another window with a few configuration queries, as shown in the following screenshot:

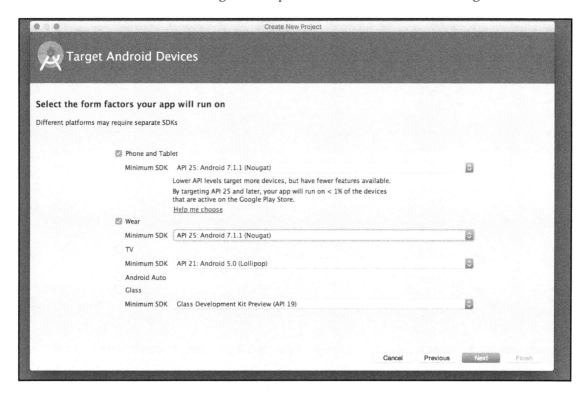

In this window, you can choose to write a standalone Wear application if you uncheck the Phone and Tablet option. In this way, you will see only Wear application templates:

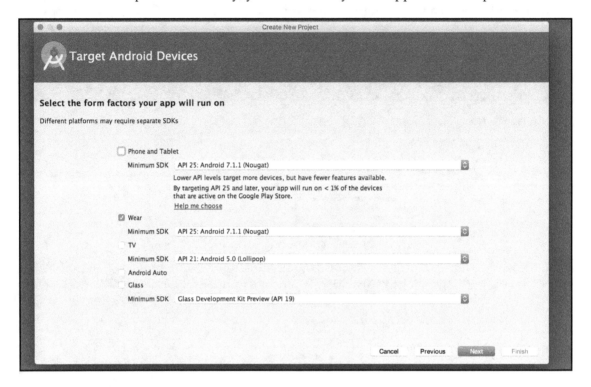

Now, Android Studio templates prompt only Android Wear activity templates with the following set of options:

- Add No Activity
- **Always On Wear Activity**
- **Blank Wear Activity**
- **Display Notification**
- **Google Maps Wear Activity**
- **Watch Face**

The activity template chooser helps you to access the default boilerplate codes, which are already templatized and can be used directly in projects:

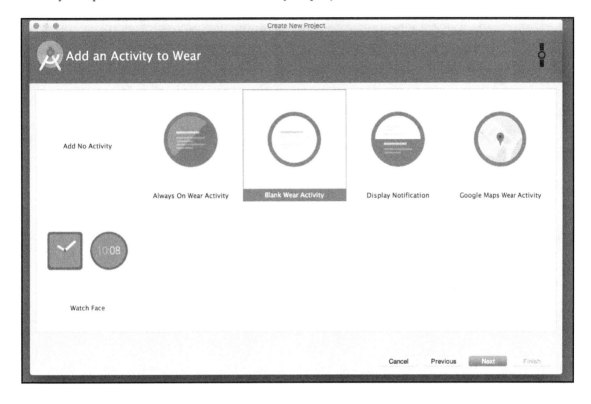

To create the first project, we will choose **Blank Wear Activity** and click on the **Next** button in the window. Android Studio will prompt another window for creating the name of the activity and layout file. In this template, the two form factors of Android Wearable devices, which are mostly round and square shapes, are prepopulated with the boilerplate code stub:

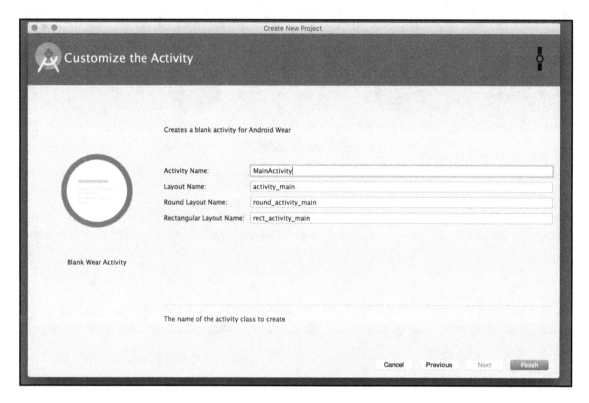

When your project is ready to be created, click on the **Finish** button. After clicking on **Finish**, Android Studio will take a few moments to create the project for us.

Way to go! You have now created a working boilerplate code for the Android Wear standalone application without the phone companion application. When successfully created, you will see the following files and codes added to your project by default:

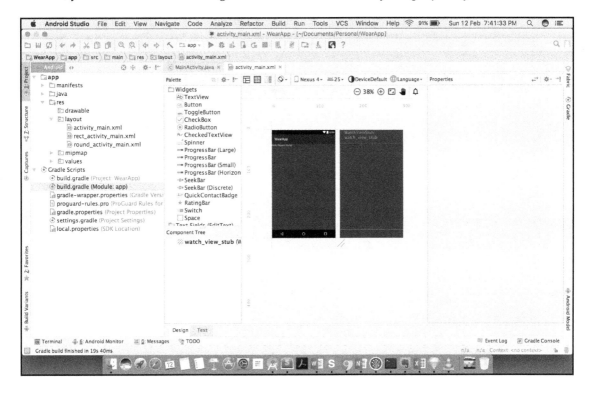

If your SDK is not updated with API level 25, you might see the Wear option in the Android Studio project creating prompts with Android Wear support library 1.x; you can update this in the Wear module Gradle file with the following dependency:

```
compile 'com.google.android.support:wearable:2.0.0'
```

Creating a Wear emulator

The process of creating a Wear emulator is very similar to creating a phone emulator.

In the AVD manager, click on the **Create Virtual Device...** button:

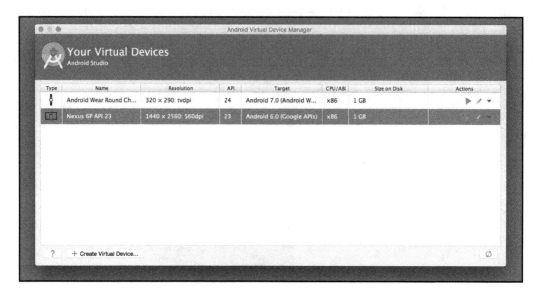

Choose the required form factor emulator according to your application needs. Now, let's create the Android Wear square emulator:

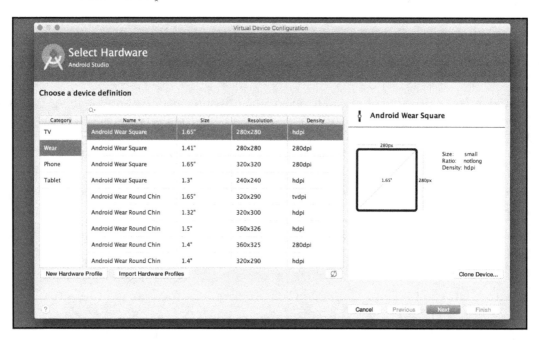

After selecting the right emulator for your Wear, you will get another prompt to choose the Wear operating system. Let's choose the **API Level 25 Nougat** emulator, as shown in the following screenshot:

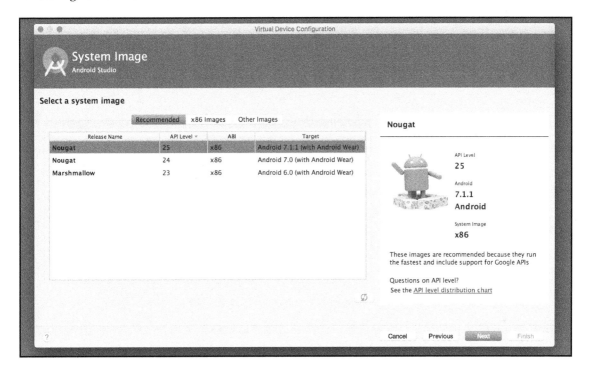

The last prompt asks for the emulator name and other orientation configurations based on your needs:

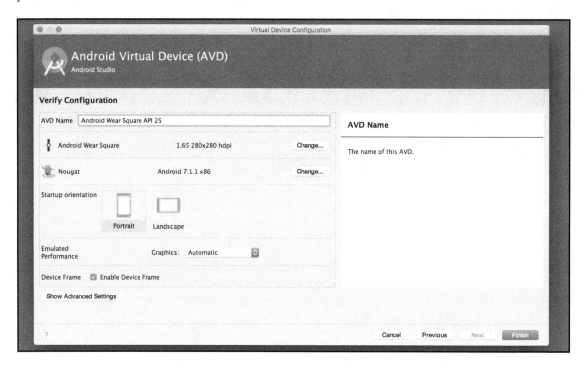

Way to go! Now, we have successfully created a square form factor emulator for the project. Let's run the project that we have created in the emulator:

Google recommends developing Wear apps in the actual hardware device to have the best user experience. However, working on emulators has the benefit of creating different screen form factors to check the application's rendering.

Working with actual Wear device

1. Open the settings menu on the Wear device
2. Go to About device
3. Click on the build number seven times to enable developer mode
4. Now enable ADB debugging on the watch

You can now connect the Wear device directly to your machine with the USB cable. You can debug your applications over Wi-Fi and Bluetooth with the following setups.

Debugging over Wi-Fi

Make sure your watch has the developer options enabled. Debugging over Wi-Fi is possible only when the Wear device and machine are connected to the same network.

- In the Wear device developer option, tap on Debug over Wi-Fi
- The watch will display its IP address (for example, 192.168.1.100). Keep a reference; we need this for the next step.
- Connect the debugger to the device
- Using the following command, we can attach the actual device to the ADB debugger:

```
adb connect 192.168.1.100
```

Enable Bluetooth debugging

We need to ensure debugging is enabled in developer options, as follows:

- Enable Debug over Bluetooth
    - Install the companion app on the phone (download it from `https://play.google.com/store/apps/details?id=com.google.android.wearable.app&hl=en`)
- Go to settings in the companion app
- Enable debugging over Bluetooth
- Connect the phone to the machine through the cable
- You can use the following commands to establish the connection:

```
adb forward tcp:4444 localabstract:/adb-hub
adb connect 127.0.0.1:4444
```

In your Android Wear, just allow ADB Debugging when it asks.

Now that we have a working setup of our development environment, let's understand the basic Android Wear-specific UI components.

Summary

In this chapter, we have looked at the initial setup for Wear application development. We have understood the necessary components to download, setting up a Wear emulator, connecting the Wear emulator to the ADB bridge, debugging over Wi-Fi, and essential user interface components specific to Wear development. In the next chapter, we will explore how to build a note-taking application which persists the data that users enter.

2
Let us Help Capture What is on Your Mind - WearRecyclerView and More

There are numerous ways to take notes. You could carry a notebook and pen in your pocket, or scribble thoughts on a piece of paper. Or, better yet, you could use your Android Wear to take notes, so one can always have a way to store thoughts even if there's not a pen and notebook nearby.

The **Note-Taking App** provides convenient access to store and retrieve notes within Android Wear devices. There are many Android smartphone note-taking apps, which are popular for their simplicity and elegant functionality. For the scope of a Wear device, it is necessary to keep the design simple and glanceable. Here, in this chapter, let's build a complete, functional, and persistent Note-Taking Wear application. We are calling this application **Wear-Note 1**:

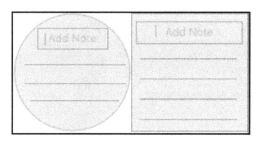

Getting started

We are going to use all the new components from Wear 2.0 API Support library. We will cover the following topics:

- `WearableRecyclerView`
- `WearableRecyclerView` adapter
- `SharedPreferences`
- `BoxInsetLayout`
- Animated `DelayedConfirmation`

Let's get started by creating a project

You learned how to create a Wear project in the previous chapter. We will be following a similar process with a slightly different configuration. In this project, we will choose Mobile and Wear app templates and extend them with further improvements in a later chapter.

Create a project called `Wear-note-1`:

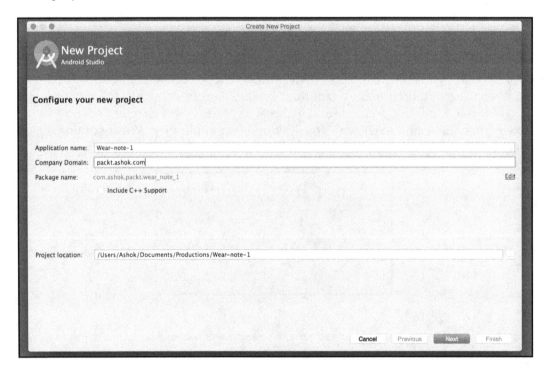

1. Your package address should be unique and should follow the reverse domain standard as Android standard development practices recommend it (for instance, `com.yourname.projectname`).
2. Select a convenient project location depending on your accessibility.
3. Select the form factors for your application to work on. Here, in this prompt, we will select both the Wear and mobile projects and select the most recent stable API 25 Nougat for the mobile and Wear projects:

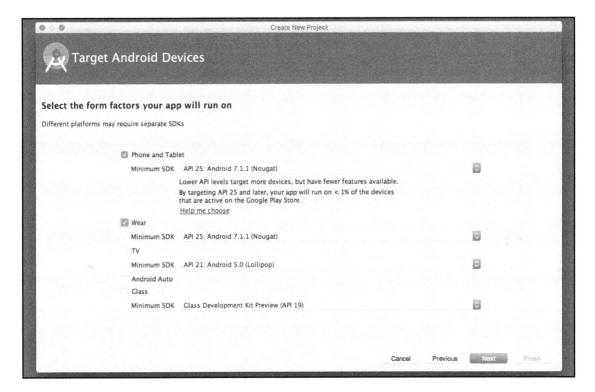

 When you select the most recent API level, your application can't be ported back to the older device; we might miss the large audience base. Seek the advantage of exploring the new features of the new API without any setbacks.

4. In the following prompt, we will be adding Activity for the mobile application; we shall choose the appropriate template from the prompt. For the scope of this project, let's select Empty Activity.

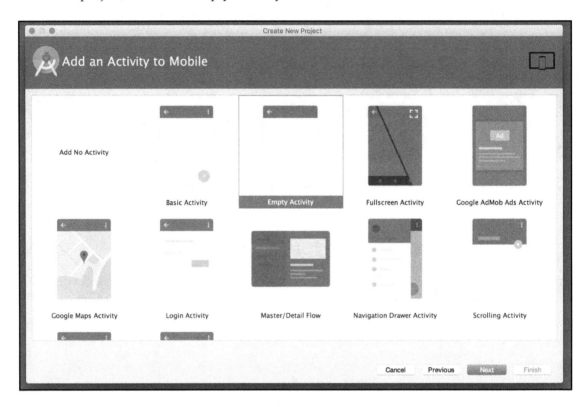

5. In the next prompt, we shall give a name for the mobile application activity.

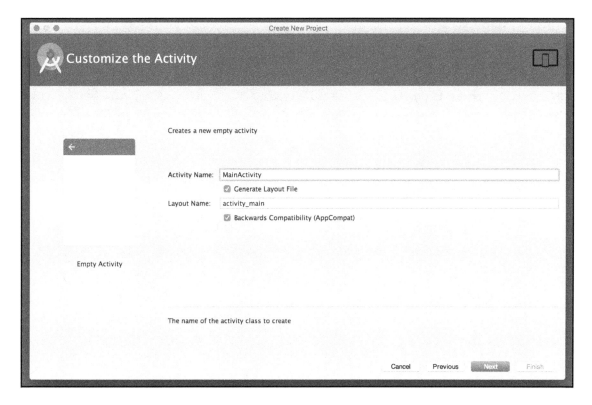

The previous prompt generates an `Hello World` application for mobile and generates all the necessary code stub.

6. This prompt allows the user to choose a different set of Wear templates. Let the default selected template be the Wear application code stub, **Always On Wear Activity**:

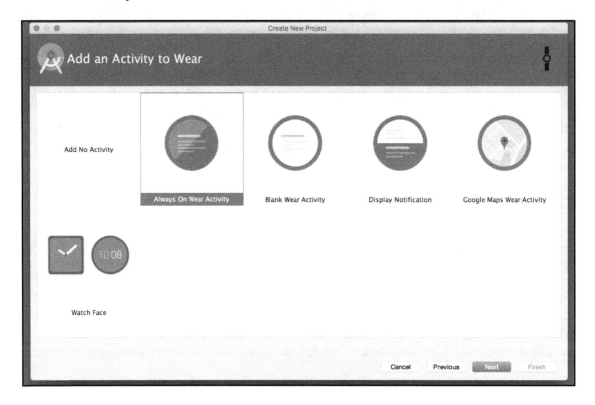

7. In the following prompt, give a suitable name for your Wear activity.

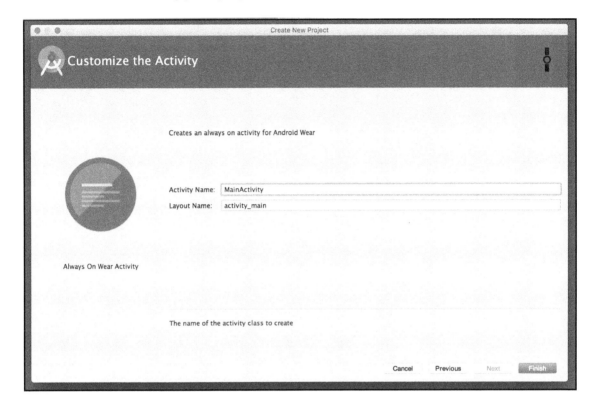

8. On successful project creation, all the necessary gradle dependencies and the Java code stubs will be created. Android Studio will have the following configuration:

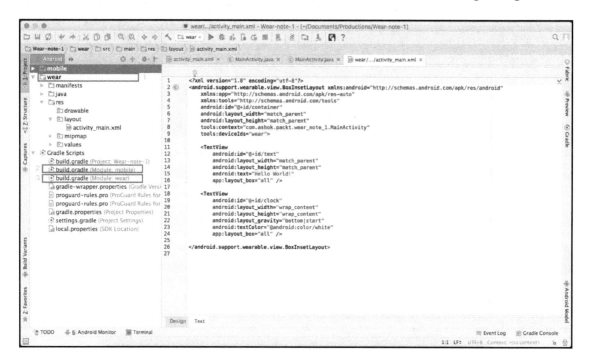

- **wear** Project
- Gradle Module: **mobile**
- Gradle Module: **wear**

The gradle module for mobile should include the Wear project in its dependency:

```
wearApp project(':wear')
```

The complete gradle module for mobile looks as follows:

```
build.gradle (Module: mobile)

apply plugin: 'com.android.application'

android {
    compileSdkVersion 25
    buildToolsVersion "25.0.2"
    defaultConfig {
        applicationId "com.ashok.packt.wear_note_1"
        minSdkVersion 25
```

```
            targetSdkVersion 25
            versionCode 1
            versionName "1.0"
            testInstrumentationRunner
            "android.support.test.runner.AndroidJUnitRunner"
        }
    buildTypes {
        release {
            minifyEnabled false
            proguardFiles getDefaultProguardFile('proguard-
            android.txt'), 'proguard-rules.pro'
        }
    }
}

dependencies {
    compile fileTree(dir: 'libs', include: ['*.jar'])
    androidTestCompile('com.android.support.test.espresso:espresso-
    core:2.2.2', {
        exclude group: 'com.android.support', module: 'support-
        annotations'
    })
    wearApp project(':wear')
    compile 'com.google.android.gms:play-services:10.0.1'
    compile 'com.android.support:appcompat-v7:25.1.1'
    testCompile 'junit:junit:4.12'
}
```

In the gradle module for Wear, we can include wear-specific gradle dependencies:

```
compile 'com.google.android.support:wearable:2.0.0'
compile 'com.google.android.gms:play-services-wearable:10.0.1'
provided 'com.google.android.wearable:wearable:2.0.0'
```

The complete gradle module for wear looks as follows:

```
apply plugin: 'com.android.application'

android {
    compileSdkVersion 25
    buildToolsVersion "25.0.2"
    defaultConfig {
        applicationId "com.ashok.packt.wear_note_1"
        minSdkVersion 25
        targetSdkVersion 25
        versionCode 1
        versionName "1.0"
    }
```

```
    buildTypes {
        release {
            minifyEnabled false
            proguardFiles getDefaultProguardFile('proguard-
            android.txt'), 'proguard-rules.pro'
        }
    }
}

dependencies {
    compile fileTree(dir: 'libs', include: ['*.jar'])
    compile 'com.google.android.support:wearable:2.0.0'
    compile 'com.google.android.gms:play-services-wearable:10.0.1'
    provided 'com.google.android.wearable:wearable:2.0.0'
}
```

Now, we are all set to get started on making a few corrections and code removal in the generated code stub for the **note** application.

> *Note: We will be working only on the Wear project in this chapter; we will not explore anything related to the mobile project. All the class creation and package creation should happen inside the Wear project.*

In the generated code stub for the Wear application, `MainActivity`, except the `onCreate()` method, removes all the other methods and codes:

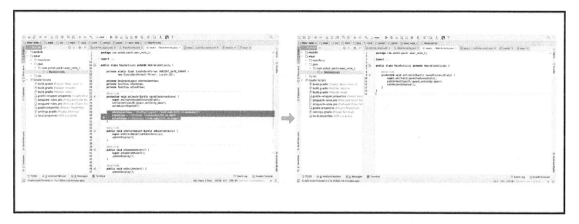

Under the `res` directory in the project section, go to the `layout` directory, select the `activity_main.xml` file, and remove the `TextView` components generated. Keep the `BoxInsetLayout` container. The complete XML will look as follows:

```
<?xml version="1.0" encoding="utf-8"?>
<android.support.wearable.view.BoxInsetLayout
```

```
xmlns:android="http://schemas.android.com/apk/res/android"
    xmlns:app="http://schemas.android.com/apk/res-auto"
    xmlns:tools="http://schemas.android.com/tools"
    android:id="@+id/container"
    android:layout_width="match_parent"
    android:layout_height="match_parent"
    tools:context="com.ashok.packt.wear_note_1.MainActivity"
    tools:deviceIds="wear">

</android.support.wearable.view.BoxInsetLayout>
```

Now, we are all set to write the note application. Before we jump into writing the code, it is very important to understand the lifecycle of an activity.

Lifecycle of an Android activity

Each screen on Android applications has a lifecycle. When you open any application, some sort of screen will welcome you. In the background, this launched screen must be created before it can convey its content to your eyes. Without you even realizing, the screen or activity experienced a couple of stages in its lifecycle. We'll discuss what lifecycle implies in Android programming.

During its lifetime, an Android activity will be in one of the following states:

- Running
- Paused
- Stopped
- Killed

The lifecycle callback methods are as follows:

- onCreate()
- onStart()
- onResume()
- onPause()
- onStop()
- onDestroy()
- onRestart()

The following diagram explains the lifecycle of an activity:

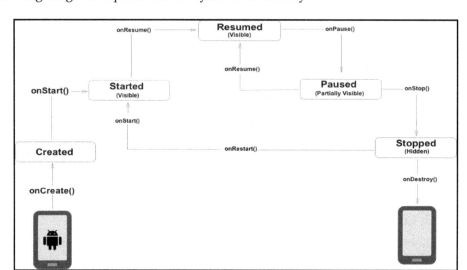

1. **onCreate()**: Basically, this event gets fired just once when the activity is started. This is a good place to initialize any information, such as, variables' lists view components and so on.
2. **onStart()**: Once the activity enters the "started" state, the activity becomes visible and interactive.
   In this method, the Android system maintains all the code that are related to the user interface and business logic of the application.
3. **onResume()**: This event is called whenever the user returns to the activity after leaving from the activity — such as receiving a call, pressing the home button, turning off the screen, or transitioning to another activity.
4. **onPause()**: The activity calls this method as the first indication that the user is leaving your activity and this method sets all the process to be idle.
5. **onStop()**: Using this callback activity is no longer visible to the user because either a new activity gets started or the existing activity gets the resumed state. The next callback that the system calls is either **onRestart()** to bring the activity back to interact with the user, or **onDestroy()** to terminate the activity.

6. **onDestroy()**: This method is called before the activity is destroyed. This is the final call that the activity receives. Using this method, we can unregister the hardware, such as accelerometer, microphone, and so on. We can terminate all the file writing operations in this method.

Now, without waiting any further, let's get started on creating packages and start coding.

Creating the packages

Creating packages helps in maintaining the code in a structured manner. In the Wear-note-1 application, we will create the following package structure:

- `model`
- `utils`
- `adapter`
- `activity`

Create these packages in the Wear application and add the activity inside the activity package. The following screenshot explains how the package looks:

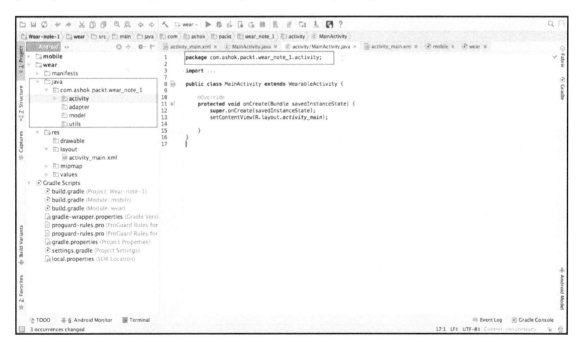

1. All the packages inside the Wear project scope.
2. Main Activity moved inside the activity package.

The conceptualized Wear note-taking app should save notes and should be able to delete notes. For persisting the notes, we shall use **SharedPreferences**, which allows all the **Create**, **Read**, **Update**, and **Delete** (CRUD) operations in terms of key-value pairs. SharedPreferences wraps inside the Android SDK; we need not worry about installing a third-party library for this chapter.

Let's create a POJO model called Note.java inside the model package in the Wear project scope. And let's create the following member variables inside the Note.java class:

 POJO stands for **Plain Old Java Object**, and would be used to describe the same things as a "Normal Class." They use getters and setters to protect their member variables.

```
private String notes = "";
private String id = "";
```

After creating the previous member variables, we shall create a parameterized constructor as follows:

```
public Note(String id, String notes) {
        this.id = id;
        this.notes = notes;
}
```

Once the constructor is completed, we need to create setters and getters for the same member variables. Android Studio helps in creating setters and getters for the requested data type using the following shortcut:

Windows/ Linux	Mac
Alt + Insert	*Command + N*

Use the previous shortcut in the **Generate...** option and select **Getter** and **Setter**. Then, choose the variables that need **Getter** and **Setter**. For this note application, we only have the id and actual note. We shall choose id and note.

The complete `Note.java` class looks as follows:

```
package com.ashok.packt.wear_note_1.model;

/**
 * Created by ashok.kumar on 15/02/17.
 */

public class Note {

    private String notes = "";
    private String id = "";

    public Note(String id, String notes) {
        this.id = id;
        this.notes = notes;
    }

    public String getNotes() {
        return notes;
    }

    public void setNotes(String notes) {
        this.notes = notes;
    }

    public String getId() {
        return id;
    }

    public void setId(String id) {
        this.id = id;
    }

}
```

Now, this POJO class will take care of setting and getting the value from and to `SharedPreference`.

Creating the SharedPreference utility

The `SharedPreference` utility class acts as an interface between the XML key value pair to read and write the data. We will create three methods for saving notes, fetching all the notes, and removing one particular note:

- `saveNote(..);`
- `getAllNotes(..);`
- `removeNote(..);`

Inside the util package, let's create two Java classes called `Constants.java` and `SharedPreferencesUtils.java`.

Inside `Constants.java`, let's add the following constants:

```
public static final String NOTE_ID = "NOTE_ID";
public static final String NOTE_TITLE = "NOTE_TITLE";
public static final String NOTE_LIST = "NOTE_LIST";
public static final String ACTION = "ACTION";
public static final String ITEM_POSITION = "ITEM_POSITION";

public static final int ACTION_ADD = 9001;
public static final int ACTION_DELETE = 9002;
public static final int ACTION_SYNC = 9003;
```

The complete class looks as follows:

```
package com.ashok.packt.wear_note_1.utils;

/**
 * Created by ashok.kumar on 15/02/17.
 */

public class Constants {

  public static final String NOTE_ID = "NOTE_ID";
  public static final String NOTE_TITLE = "NOTE_TITLE";
  public static final String NOTE_LIST = "NOTE_LIST";
  public static final String ACTION = "ACTION";
  public static final String ITEM_POSITION = "ITEM_POSITION";

  public static final int ACTION_ADD = 9001;
  public static final int ACTION_DELETE = 9002;
  public static final int ACTION_SYNC = 9003;

}
```

Now, it's the time to work on the `SharedPreferencesUtils.java` class. In this class, we will create the methods to save data, retrieve data, and delete data.

Saving notes

The saving note method accepts two arguments, `Note` and `Context`. When a note is not empty, it opens up the `SharedPreference` editor and saves the data:

```
public static void saveNote(Note note, Context context) {
    if (note != null) {
        SharedPreferences sharedPreferences =
        PreferenceManager.getDefaultSharedPreferences(context);
        SharedPreferences.Editor editor = sharedPreferences.edit();
        editor.putString(note.getId(), note.getNotes());
        editor.apply();
    }
}
```

Fetching all the saved notes from SharedPreference

Fetching all the notes for this List type method needs a Context and then the `SharedPreference` reference using the `PreferenceManager` class. Then, we will create an instance of a List of notes for adding the notes after fetching it. We use the Map type for looping through the saved data inside `SharedPreference`. We will add all the data to the list inside the loop; it returns the `noteList`:

```
public static List<Note> getAllNotes(Context context) {
    SharedPreferences sharedPreferences =
    PreferenceManager.getDefaultSharedPreferences(context);
    List<Note> noteList = new ArrayList<>();
    Map<String, ?> key = sharedPreferences.getAll();
    for (Map.Entry<String, ?> entry : key.entrySet()) {
        String savedValue = (String) entry.getValue();

        if (savedValue != null) {
            Note note = new Note(entry.getKey(), savedValue);
            noteList.add(note);
        }
    }
    return noteList;
}
```

Removing notes from SharedPreference

To remove the data from `SharedPreference`, we need the key for the particular data item that we already set as the note ID. To remove the note, we need to pass the ID of the note and context. Using the `SharedPreference` editor, we will be able to remove the data:

```java
public static void removeNote(String id, Context context) {
    if (id != null) {
        SharedPreferences sharedPreferences =
        PreferenceManager.getDefaultSharedPreferences(context);
        SharedPreferences.Editor editor = sharedPreferences.edit();
        editor.remove(id);
        editor.apply();
    }
}
```

The complete class looks as follows:

```java
package com.ashok.packt.wear_note_1.utils;

import android.content.Context;
import android.content.SharedPreferences;
import android.preference.PreferenceManager;

import com.ashok.packt.wear_note_1.model.Note;

import java.util.ArrayList;
import java.util.List;
import java.util.Map;

/**
 * Created by ashok.kumar on 15/02/17.
 */

public class SharedPreferencesUtils {

    public static void saveNote(Note note, Context context) {
        if (note != null) {
            SharedPreferences sharedPreferences =
            PreferenceManager.getDefaultSharedPreferences(context);
            SharedPreferences.Editor editor =
            sharedPreferences.edit();
            editor.putString(note.getId(), note.getNotes());
            editor.apply();
        }
    }
```

```java
public static List<Note> getAllNotes(Context context) {
    SharedPreferences sharedPreferences =
    PreferenceManager.getDefaultSharedPreferences(context);
    List<Note> noteList = new ArrayList<>();
    Map<String, ?> key = sharedPreferences.getAll();
    for (Map.Entry<String, ?> entry : key.entrySet()) {
        String savedValue = (String) entry.getValue();

        if (savedValue != null) {
            Note note = new Note(entry.getKey(), savedValue);
            noteList.add(note);
        }
    }
    return noteList;
}

public static void removeNote(String id, Context context) {
    if (id != null) {
        SharedPreferences sharedPreferences =
        PreferenceManager.getDefaultSharedPreferences(context);
        SharedPreferences.Editor editor =
        sharedPreferences.edit();
        editor.remove(id);
        editor.apply();
    }
}
}
```

Building the User Interface

The **User Interface (UI)** needs to be simple and glanceable. For managing different form factors of a Wear device, such as a circle and square, we will be using a container, which is also referred to as a screen shape-aware frame layout. For getting the best listing experience, we will be using the latest component, `WearableRecyclerView`, from the Wear 2.0 support library. `WearableRecyclerView` has the best architecture, which is identical to `RecyclerView`.

For `activity_main.xml`, add the following code:

```xml
<?xml version="1.0" encoding="utf-8"?>
<android.support.wearable.view.BoxInsetLayout
    xmlns:android="http://schemas.android.com/apk/res/android"
    xmlns:tools="http://schemas.android.com/tools"
    xmlns:app="http://schemas.android.com/apk/res-auto"
    android:layout_width="match_parent"
    android:layout_height="match_parent"
```

```
        android:id="@+id/container"
        tools:context="com.ashok.packt.wear_note_1.activity.MainActivity"
        tools:deviceIds="wear"
        app:layout_box="all"
        android:padding="5dp">

        <LinearLayout android:layout_width="match_parent"
            android:layout_height="match_parent"
            android:orientation="vertical"
            android:layout_gravity="center">

            <EditText
                android:layout_width="match_parent"
                android:layout_height="50dp"
                android:id="@+id/edit_text"
                android:gravity="center"
                android:inputType="textCapSentences|textAutoCorrect"
                android:layout_gravity="center"
                android:imeOptions="actionSend"
                android:textColorHint="@color/orange"
                android:hint="@string/add_a_note"/>

            <android.support.wearable.view.WearableRecyclerView
                android:id="@+id/wearable_recycler_view"
                android:layout_width="match_parent"
                android:layout_height="match_parent"/>
        </LinearLayout>
</android.support.wearable.view.BoxInsetLayout>
```

 Note: In case you neglect to include the necessary strings in the `strings.xml` file, you will experience string missing errors in the activity layout files. Include the strings mandatorily, which we will see now.

Before adding all the UI components, let's add all the string resources to be used. This resource is assumed for the operation that we have planned to implement in the Wear note app.

Add the following resources in `strings.xml` under the `res` directory and the `value` directory:

```
<resources>
    <string name="app_name">Wear-note-1</string>
    <string name="hello_world">Hello World!</string>
    <string name="add_a_note">Add a Note</string>
    <string name="delete">Remove</string>
    <string name="note_saved">Saved</string>
```

```xml
    <string name="note_removed">Note Removed</string>
    <string name="cancel">Cancelled</string>
</resources>
```

Next, let's add the `WearableRecyclerView` list item layout. It's called `each_item.xml`. For creating a new layout, go to the `res` directory and navigate to the `layout` directory. Right-click and go to **New** and then select the **Layout resource file**. Once the layout file is created, add the following code:

```xml
<?xml version="1.0" encoding="utf-8"?>
<LinearLayout
    xmlns:android="http://schemas.android.com/apk/res/android"
    xmlns:tools="http://schemas.android.com/tools"
    android:layout_width="match_parent"
    android:layout_height="match_parent"
    android:gravity="center"
    android:layout_gravity="center"
    android:clickable="true"
    android:background="?android:attr/selectableItemBackground"
    android:orientation="vertical">

    <TextView
        android:id="@+id/note"
        android:layout_width="wrap_content"
        android:layout_height="wrap_content"
        android:gravity="center"
        android:layout_gravity="center"
        tools:text="note"/>
    <View
        android:layout_width="match_parent"
        android:layout_height="1px"
        android:layout_marginLeft="5dp"
        android:layout_marginRight="5dp"
        android:background="@android:color/darker_gray"/>
</LinearLayout>
```

In API level 21, Google officially introduced the `RecyclerView` component. `RecyclerView` is a flexible and efficient version of `ListView`. It is a container for rendering a larger dataset of views that can be recycled and scrolled very efficiently. `RecyclerView` is like a traditional `ListView` component but with more flexibility to customize the rendering and performance. `WearableRecyclerview` is exclusively customized for wearable devices.

Creating an adapter for WearableRecyclerView

An adapter manages the data model and adapts it to each row in `RecyclerView` or `ListView`. The adapter handles the filtering and sorting:

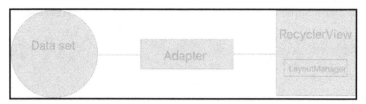

Adapter implementation

Create a class called `RecyclerViewAdapter` in the `adapter` package created earlier. Inside the class, let's create an inner class called `ViewHolder` for mapping the components created in the `each_item.xml` layout. `ViewHolder` will extend `RecyclerView.ViewHolder` and, inside its constructor, it will map the XML component. The code looks as follows:

```
public class RecyclerViewAdapter {

    static class ViewHolder extends RecyclerView.ViewHolder{

        private TextView mTextView;

        public ViewHolder(View view) {
            super(view);
            mTextView = (TextView) view.findViewById(R.id.note);
        }
    }
}
```

Now, create an interface for the `ItemClick` events, which returns the position:

```
public interface ItemSelectedListener {
    void onItemSelected(int position);
}
```

Inside the `ViewHolder` class, add the bind method for binding the click event and returning the position:

```
void bind(final String title, final int position, final
ItemSelectedListener listener) {

    mTextView.setText(title);
    itemView.setOnClickListener(new View.OnClickListener() {
        @Override
```

```
    public void onClick(View v) {
        if (listener != null) {
            listener.onItemSelected(position);
        }
    }
    });
}
```

In the `RecyclerViewAdapter` class scope, declare the following instances:

```
private ItemSelectedListener mItemSelectedListener;
private List<Note> mListNote = new ArrayList<>();
```

Set the listner for `ItemSelectedListner` as follows:

```
public void setListener(ItemSelectedListener itemSelectedListener) {
    mItemSelectedListener = itemSelectedListener;
}
```

Now, we need to extend `RecyclerViewAdapter` to `WearableRecyclerView.Adapter` with its view holder, as follows:

```
public class RecyclerViewAdapter extends
WearableRecyclerView.Adapter<RecyclerViewAdapter.ViewHolder> {

}
```

We then need to implement the callbacks for the adapter:

- `onCreateViewHolder(..);`
- `onBindViewHolder(..);`
- `getItemCount(..);`

Inside `onCreateViewHolder`, we should attach or inflate the `each_item.xml` layout:

```
@Override
public RecyclerViewAdapter.ViewHolder onCreateViewHolder(ViewGroup
parent, int viewType) {

View view = LayoutInflater.from(parent.getContext())
.inflate(R.layout.view_holder, parent, false);

return new ViewHolder(view);
}
```

Inside `onBindViewHolder`, we will handle the click event by binding the data and position:

```
@Override
public void onBindViewHolder(ViewHolder holder, int position) {
    holder.bind(mListNote.get(position).getNotes(), position,
    mItemSelectedListener);
}
```

Inside the `getItemCount()` method, we will return the count of the list item being passed to the adapter:

```
@Override
public int getItemCount() {
    return mListNote.size();
}
```

The complete `WearableRecyclerView` adapter looks as follows:

```
package com.ashok.packt.wear_note_1.adapter;

import android.support.v7.widget.RecyclerView;
import android.support.wearable.view.WearableRecyclerView;
import android.view.LayoutInflater;
import android.view.View;
import android.view.ViewGroup;
import android.widget.TextView;

import com.ashok.packt.wear_note_1.R;
import com.ashok.packt.wear_note_1.model.Note;

import java.util.ArrayList;
import java.util.List;

/**
 * Created by ashok.kumar on 15/02/17.
 */

public class RecyclerViewAdapter
        extends
WearableRecyclerView.Adapter<RecyclerViewAdapter.ViewHolder> {

    private ItemSelectedListener mItemSelectedListener;
    private List<Note> mListNote = new ArrayList<>();

    static class ViewHolder extends RecyclerView.ViewHolder {
        private TextView mTextView;

        ViewHolder(View view) {
```

```
            super(view);
            mTextView = (TextView) view.findViewById(R.id.note);
        }

        void bind(final String title, final int position, final
        ItemSelectedListener listener) {
            mTextView.setText(title);
            itemView.setOnClickListener(new View.OnClickListener() {
                @Override
                public void onClick(View v) {
                    if (listener != null) {
                        listener.onItemSelected(position);
                    }
                }
            });
        }
    }

    public void setListener(ItemSelectedListener itemSelectedListener)
    {
        mItemSelectedListener = itemSelectedListener;
    }

    public void setListNote(List<Note> listNote) {
        mListNote.clear();
        mListNote.addAll(listNote);
        notifyDataSetChanged();
    }

    @Override
    public RecyclerViewAdapter.ViewHolder onCreateViewHolder(ViewGroup
    parent, int viewType) {
        View view = LayoutInflater.from(parent.getContext())
                .inflate(R.layout.each_item, parent, false);

        return new ViewHolder(view);
    }

    @Override
    public void onBindViewHolder(ViewHolder holder, int position) {
        holder.bind(mListNote.get(position).getNotes(), position,
        mItemSelectedListener);
    }

    @Override
    public int getItemCount() {
        return mListNote.size();
    }
```

```
        public interface ItemSelectedListener {
            void onItemSelected(int position);
        }
    }
```

Now, let's put all these pieces of codes for action in `MainActivity.java`.

Working on activities and driving the project towards completion

In `MainActivity.java`, we will have control over the classes that we have worked on so far. Let's get started on implementing all the necessary code and methods.

When we extend the Wear project activity to `WearableActivity`, we shall include the following code in the manifest `<application ></application>` tag scope: `<uses-library android:name="com.google.android.wearable" android:required="false" />`. When we declare `android: required="false"` any feature that Android has control on will be disabled. In the context of Wear applications, when you need to extend your activity with `WearableActivity`, we need to set it to false.

Now, in `MainActivity.java`, let's add the basic variable instances necessary for the project. Add the following code globally in the `MainActivity.java` class scope:

```
private static final String TAG = "MainActivity";
private static final int REQUEST_CODE = 1001;
private RecyclerViewAdapter mAdapter;
private List<Note> myDataSet = new ArrayList<>();
```

Next, write a method for configuring the UI component. Let's call this method `ConfigureUI()`. In this method, we shall handle the click event. When a user clicks on the `RecyclerView` list item, it should take the user to the delete activity screen:

```
private void configureUI() {
    WearableRecyclerView recyclerView = (WearableRecyclerView)
    findViewById(R.id.wearable_recycler_view);
    recyclerView.setHasFixedSize(true);
    LinearLayoutManager mLayoutManager = new
    LinearLayoutManager(this);
    recyclerView.setLayoutManager(mLayoutManager);

    mAdapter = new RecyclerViewAdapter();
    mAdapter.setListNote(myDataSet);
    mAdapter.setListener(this);
    recyclerView.setAdapter(mAdapter);
```

```
EditText editText = (EditText) findViewById(R.id.edit_text);

editText.setOnEditorActionListener(new
TextView.OnEditorActionListener() {
    @Override
    public boolean onEditorAction(TextView textView, int
    action, KeyEvent keyEvent) {
        if (action == EditorInfo.IME_ACTION_SEND) {
            String text = textView.getText().toString();
            if (!TextUtils.isEmpty(text)) {
             //Todo add the on editText click handlers
             return true;
            }
        }
        return false;
    }
});
}
```

Writing a method to create a note when clicking on Add a Note

The `CreateNote` method accepts two arguments, `id` and `note`. If `id` is null, it returns the current system in a millisecond:

```
private Note createNote(String id, String note) {
    if (id == null) {
        id = String.valueOf(System.currentTimeMillis());
    }
    return new Note(id, note);
}
```

Adding a method for updating the adapter

The `updateAdapter` method clears the dataset, requests `SharedPreferencesUtils` to fetch the all the notes, and sets it to the adapter:

```
private void updateAdapter() {
    myDataSet.clear();
    myDataSet.addAll(SharedPreferencesUtils.getAllNotes(this));
    mAdapter.setListNote(myDataSet);
}
```

Now, before we continue to add a method for updating the data, it might be a good idea to add the `ConfirmationUtil.java` class in the `utils` directory for helping the delayed confirmation and other Wear animations. Once, after creating the `ConfirmationUtil.java` class, add the following show message method with `ConfirmationActivity` provided in the wearable support library:

```
public class ConfirmationUtils {
    public static void showMessage(String message, Context context) {
        Intent intent = new Intent(context,
        ConfirmationActivity.class);
        intent.putExtra(ConfirmationActivity.EXTRA_ANIMATION_TYPE,
        ConfirmationActivity.SUCCESS_ANIMATION);
        intent.putExtra(ConfirmationActivity.EXTRA_MESSAGE, message);
        context.startActivity(intent);
    }
}
```

Next, add the `updateData` method.

Adding the updateData method

In the update data method, we will check whether the action is for adding a note or for deleting the note. We will fetch the constants from the util constants. After adding the notes, we will show a success message added in the string resources already. The method looks as follows:

```
private void updateData(Note note, int action) {
    if (action == Constants.ACTION_ADD) {
        SharedPreferencesUtils.saveNote(note, this);
        ConfirmationUtils.showMessage(getString
        (R.string.note_saved), this);
    } else if (action == Constants.ACTION_DELETE) {
        SharedPreferencesUtils.removeNote(note.getId(), this);
        ConfirmationUtils.showMessage(getString
        (R.string.note_removed), this);
    }
    updateAdapter();
}
```

Preparing the method for updating

In the prepareUpdate method, If the data is not empty, we shall update it:

```
private void prepareUpdate(String id, String title, int action) {
    if (!(TextUtils.isEmpty(id) && TextUtils.isEmpty(title))) {
        Note note = createNote(id, title);
        updateData(note, action);
    }
}
```

Updating the data on the UI when the application restarts

When an app loses its app state in the onResume method, we shall update the user interface with all the persisted data in SharedPreferences:

```
@Override
protected void onResume() {
    super.onResume();
    updateAdapter();
}
```

Now, in onItemClickListner, we should take the user to deleteActivity for the note deletion. When we start the intent, we should use startActivityForResult(), as follows. We will get requestcode and responsecode in the onActivityResult method for handling the action:

```
@Override
public void onItemSelected(int position) {
    Intent intent = new Intent(getApplicationContext(),
    DeleteActivity.class);
    intent.putExtra(Constants.ITEM_POSITION, position);
    startActivityForResult(intent, REQUEST_CODE);
}
```

Overriding onActivityResult for getting the delete reference

When a user clicks on the list item, he or she will be navigated to the `deleteConfirmationView` activity for deleting the item. On successful operation, the data item will get updated on the screen:

```
@Override
protected void onActivityResult(int requestCode, int resultCode, Intent
data) {
    if (data != null && requestCode == REQUEST_CODE &&
    resultCode == RESULT_OK) {
        if (data.hasExtra(Constants.ITEM_POSITION)) {
            int position = data.getIntExtra
            (Constants.ITEM_POSITION, -1);
            if (position > -1) {
                Note note = myDataSet.get(position);
                updateData(note, Constants.ACTION_DELETE);
            }
        }
    }
}
```

The complete `MainActivity.java` class looks as follows:

```
package com.ashok.packt.wear_note_1.activity;

import android.content.Intent;
import android.os.Bundle;
import android.support.v7.widget.LinearLayoutManager;
import android.support.wearable.activity.WearableActivity;
import android.support.wearable.view.WearableRecyclerView;
import android.text.TextUtils;
import android.view.KeyEvent;
import android.view.inputmethod.EditorInfo;
import android.widget.EditText;
import android.widget.TextView;

import com.ashok.packt.wear_note_1.R;
import com.ashok.packt.wear_note_1.adapter.RecyclerViewAdapter;
import com.ashok.packt.wear_note_1.model.Note;
import com.ashok.packt.wear_note_1.utils.ConfirmationUtils;
import com.ashok.packt.wear_note_1.utils.Constants;
import com.ashok.packt.wear_note_1.utils.SharedPreferencesUtils;

import java.util.ArrayList;
import java.util.List;
```

```
public class MainActivity extends WearableActivity implements
RecyclerViewAdapter.ItemSelectedListener {

    private static final String TAG = "MainActivity";
    private static final int REQUEST_CODE = 1001;
    private RecyclerViewAdapter mAdapter;
    private List<Note> myDataSet = new ArrayList<>();

    @Override
    protected void onCreate(Bundle savedInstanceState) {
        super.onCreate(savedInstanceState);
        setContentView(R.layout.activity_main);
        configureUI();
    }

    private void configureUI() {
        WearableRecyclerView recyclerView = (WearableRecyclerView)
        findViewById(R.id.wearable_recycler_view);
        recyclerView.setHasFixedSize(true);
        LinearLayoutManager mLayoutManager = new
        LinearLayoutManager(this);
        recyclerView.setLayoutManager(mLayoutManager);

        mAdapter = new RecyclerViewAdapter();
        mAdapter.setListNote(myDataSet);
        mAdapter.setListener(this);
        recyclerView.setAdapter(mAdapter);

        EditText editText = (EditText) findViewById(R.id.edit_text);

        editText.setOnEditorActionListener(new
        TextView.OnEditorActionListener() {
            @Override
            public boolean onEditorAction(TextView textView, int
            action, KeyEvent keyEvent) {
                if (action == EditorInfo.IME_ACTION_SEND) {
                    String text = textView.getText().toString();
                    if (!TextUtils.isEmpty(text)) {
                        Note note = createNote(null, text);
                        SharedPreferencesUtils.saveNote(note,
                        textView.getContext());
                        updateData(note, Constants.ACTION_ADD);
                        textView.setText("");
                        return true;
                    }
                }
                return false;
            }
```

```
        });
    }

    private void updateAdapter() {
        myDataSet.clear();
        myDataSet.addAll(SharedPreferencesUtils.getAllNotes(this));
        mAdapter.setListNote(myDataSet);
    }

    @Override
    public void onItemSelected(int position) {
        Intent intent = new Intent
        (getApplicationContext(), DeleteActivity.class);
        intent.putExtra(Constants.ITEM_POSITION, position);
        startActivityForResult(intent, REQUEST_CODE);
    }

    private void updateData(Note note, int action) {
        if (action == Constants.ACTION_ADD) {
            SharedPreferencesUtils.saveNote(note, this);
            ConfirmationUtils.showMessage(getString
            (R.string.note_saved), this);
        } else if (action == Constants.ACTION_DELETE) {
            SharedPreferencesUtils.removeNote(note.getId(), this);
            ConfirmationUtils.showMessage(getString
            (R.string.note_removed), this);
        }
        updateAdapter();
    }

    private void prepareUpdate(String id, String title, int action) {
        if (!(TextUtils.isEmpty(id) && TextUtils.isEmpty(title))) {
            Note note = createNote(id, title);
            updateData(note, action);
        }
    }

    private Note createNote(String id, String note) {
        if (id == null) {
            id = String.valueOf(System.currentTimeMillis());
        }
        return new Note(id, note);
    }

    @Override
    protected void onActivityResult(int requestCode, int resultCode,
    Intent data) {
        if (data != null && requestCode == REQUEST_CODE &&
```

```
        resultCode == RESULT_OK) {
            if (data.hasExtra(Constants.ITEM_POSITION)) {
                int position =
                data.getIntExtra(Constants.ITEM_POSITION, -1);
                if (position > -1) {
                    Note note = myDataSet.get(position);
                    updateData(note, Constants.ACTION_DELETE);
                }
            }
        }
    }

    @Override
    protected void onResume() {
        super.onResume();
        updateAdapter();
    }
}
```

Let's see DeleteActivity in detail

DeleteActivity implements the
DelayedConfirmationView.DelayedConfirmationListener interface with two
callbacks to handle the ConfirmationView animation. DelayedConfirmationView can
automatically confirm the operation once the time elapses. The intended delay is to cancel
the operation:

```
public class DeleteActivity extends WearableActivity implements
DelayedConfirmationView.DelayedConfirmationListener  {



  @Override
  public void onTimerFinished(View view) {

  }

  @Override
  public void onTimerSelected(View view) {

  }

}
```

In `activity_delete.xml`, add `delayedConfirmationview` from the wearable support library. Add it inside the `BoxInsetLayout` container as follows:

```
<?xml version="1.0" encoding="utf-8"?>
<android.support.wearable.view.BoxInsetLayout
    xmlns:android="http://schemas.android.com/apk/res/android"
    xmlns:tools="http://schemas.android.com/tools"
    xmlns:app="http://schemas.android.com/apk/res-auto"
    android:layout_width="match_parent"
    android:layout_height="match_parent"
    android:id="@+id/container"
    tools:context="com.ashok.packt.wear_note_1.activity.DeleteActivity"
    tools:deviceIds="wear">

    <LinearLayout android:layout_width="match_parent"
        android:layout_height="match_parent"
        android:orientation="vertical"
        android:padding="5dp"
        app:layout_box="all"
        android:layout_gravity="center"
        android:gravity="center">

        <TextView
            android:layout_width="wrap_content"
            android:layout_height="wrap_content"
            android:text="@string/delete"/>

        <android.support.wearable.view.DelayedConfirmationView
            android:id="@+id/delayed_confirmation"
            android:layout_width="wrap_content"
            android:layout_height="wrap_content"
            android:src="@android:drawable/ic_delete"
            app:circle_border_color="@color/orange"
            app:circle_color="@color/white"
            app:circle_border_width="8dp"
            app:circle_radius="30dp"/>

    </LinearLayout>
</android.support.wearable.view.BoxInsetLayout>
```

Declare a private instance of `DelayedConfirmationView` in the `deleteActivity` scope:

```
private DelayedConfirmationView mDelayedConfirmationView;
```

Now, in the `onCreate` method, map the components with the following configuration:

```
mDelayedConfirmationView = (DelayedConfirmationView)
findViewById(R.id.delayed_confirmation);
```

```
mDelayedConfirmationView.setListener(this);
mDelayedConfirmationView.setTotalTimeMs(3000);
mDelayedConfirmationView.start();
```

Now, for the automatic operation confirmation, after the time elapse, we shall add the following code in the onTimerfinished method:

```
@Override
 public void onTimerFinished(View view) {
     int itemPosition =
     getIntent().getIntExtra(Constants.ITEM_POSITION, -1);
     Intent intent = new Intent();
     intent.putExtra(Constants.ITEM_POSITION, itemPosition);
     setResult(RESULT_OK, intent);
     finish();
 }
```

For cancelling the operation in case a user clicks on the timer or selects the timer, we need to cancel the deletion. In the onTimerSelected() method, add the following code snippet:

```
@Override
public void onTimerSelected(View view) {
    ConfirmationUtils.showMessage(getString(R.string.cancel), this);
    mDelayedConfirmationView.reset();
    finish();
}
```

Complete the DeleteActivity code as follows:

```
package com.ashok.packt.wear_note_1.activity;

import android.content.Intent;
import android.os.Bundle;
import android.support.wearable.activity.WearableActivity;
import android.support.wearable.view.DelayedConfirmationView;
import android.view.View;

import com.ashok.packt.wear_note_1.R;
import com.ashok.packt.wear_note_1.utils.ConfirmationUtils;
import com.ashok.packt.wear_note_1.utils.Constants;

public class DeleteActivity extends WearableActivity implements
DelayedConfirmationView.DelayedConfirmationListener  {

    private DelayedConfirmationView mDelayedConfirmationView;

    @Override
    protected void onCreate(Bundle savedInstanceState) {
```

```
        super.onCreate(savedInstanceState);
        setContentView(R.layout.activity_delete);
        mDelayedConfirmationView = (DelayedConfirmationView)
        findViewById(R.id.delayed_confirmation);
        mDelayedConfirmationView.setListener(this);
        mDelayedConfirmationView.setTotalTimeMs(3000);
        mDelayedConfirmationView.start();

    }

    @Override
    public void onTimerFinished(View view) {
        int itemPosition =
        getIntent().getIntExtra(Constants.ITEM_POSITION, -1);
        Intent intent = new Intent();
        intent.putExtra(Constants.ITEM_POSITION, itemPosition);
        setResult(RESULT_OK, intent);
        finish();
    }

    @Override
    public void onTimerSelected(View view) {
        ConfirmationUtils.showMessage(getString
        (R.string.cancel), this);
        mDelayedConfirmationView.reset();
        finish();
    }
}
}
```

Now, we have a complete, working note application for our Wear device, which looks as follows. All the notes are persistently saved in Android Wear.

Note application in a round form factor device

The home screen when the application launches and lands on this screen looks as follows:

When a user clicks on the note, it will take users to this screen, which dictates the deletion process based on the delay:

The Wear input method framework that allows users to type in and swipe in information looks like this:

When notes are successfully saved, the confirmation activity shows the following animation:

The delayed confirmation view shows the successful deletion animation once the time interval is completed:

When a user cancels the deletion before the time interval, the following animation will take place:

Note application in a square form factor device

The home screen in a square screen form factor looks as follows:

When a user clicks on the note, it will take the user to this screen, which dictates the deletion process based on the delay:

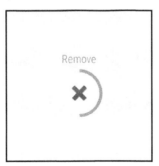

The Wear input method framework allows users to type in and swipe in information:

When notes are successfully saved, `ConfirmationActivity` shows the following animation:

The delayed confirmation view shows the successful deletion animation once the time interval is completed:

When users cancel the deletion before the time interval, the following animation will take place:

Summary

In this chapter, we have explored `WearablerecyclerView`, `DelayedConfirmationView`, and writing organized code using packages. We have explored the callbacks for all the interface that we implemented in the project. It offered a clear understanding on `SharedPreferences` using CRUD operations, working with POJO classes, working with the `RecyclerView` adapter and click events, and the necessary configuration for the Android Wear project.

3
Let us Help Capture What is on Your Mind - Saving Data and Customizing the UI

Building the note-taking application from scratch was a good learning experience. In this chapter, we will understand upgrading the same code for a new user interface with Wear design standards. The **Wear-note 1** app will get migrated to **Wear-note 2** with the following update:

- Integrating the Realm database
- UI updates
- Integrating custom fonts and assets
- Finalizing the project

To further assist the open wear-note 1 project in your Android Studio, compile the project and check the project screen by screen. You will discover that the main functionality of the application is saving notes on Wear devices. In this application, we have a white color background and black color font. `SharedPreferences` is helping the application to persist the data.

Further recapping, we know how to work with `WearableRecyclerView`, `DelayedConfirmationView`, and `BoxInsetLayout` for getting the best application experience on Wear devices.

Taking it forward, let's get started on finishing the project with the changes mentioned earlier. Let's call this application wear-note-2. Go to your `values` directory in the `res` directory, and in the `string.xml` file, change the application name to Wear-note-2, as follows:

```
<string name="app_name">Wear-note-2</string>
```

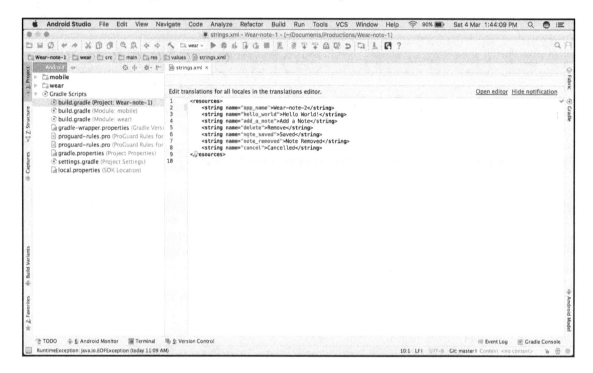

Wear-note-2

Let's start working on functionality and replacing the database with RealmDB in this submodule.

RealmDB is creating buzz in the modern Android programming world; it is a simple alternative to the SQLite database, which comes built into the Android SDK. Realm is not using SQLite as its core engine; it has its own C++ core, exclusively architected for mobiles. Realm saves the data in a universal, table-based format with a C++ core. Realm handles a range of sophisticated queries and can allow data accessibility from multiple languages as well as numerous ad hoc queries.

Advantages of Realm

- Realm is 10x faster than SQLite
- Easy to use
- Cross-platform
- Memory efficient
- Well documented and excellent community support

Exploring more about Realm, there are numerous optimizations Realm does, such as integer packing and converting common strings into enums, which results in better performance than the SQLite + ORM database solutions. The traditional SQLite + ORM abstraction is leaky, ORM simply converts objects and their methods into SQL statements, which results in poor performance. On the other hand, Realm is an object-oriented database, which means your data is saved as objects and this is reflected in your database. Realm maps the whole data in the memory using B+ trees and, whenever data is queried, Realm calculates the offset, reads from the memory mapped region, and returns the raw value. By doing this, Realm avoids zero-copy (the traditional way of reading data from a database leads to unnecessary copying (raw data > deserialized representation > language-level objects)).

Realm is the perfect choice whenever a developer wants to implement lazy-loading; because the properties are represented in columns instead of rows, it can lazy load the properties as necessary, and because of the column structure, reads are much faster, while inserts are slower. But that is a good trade-off for the context of a mobile application.

Realm uses the **Multiversion Concurrency Control** (**MVCC**) model, which means that multiple read transactions can be done at the same time and reads can also be done while a write transaction is being committed.

 For more information, visit `https://realm.io/news/jp-simard-realm-core-database-engine/`.

Disadvantages of Realm

Realm has a few bottlenecks that should be considered before choosing it. Nevertheless, these bottlenecks can be considered for a high scaling Android app:

- We cannot import Realmdb to other applications
- We can't access the objects across threads
- Doesn't have auto-increment of IDs
- Migrating Realmdb is a painful job
- Lacks compound primary key
- Still under active development

In the **Wear-note-1** project, after changing its name in `string.xml` to Wear-note-2, we need to add the appropriate Realm dependency in the gradle-project module. As of the time of writing this book, the newest Realm version is 3.0.0, so let's discuss the dependency and code in detail.

Re-structuring the code and dependencies

Realm has a new mechanism now. It's not just a gradle dependency, it's a gradle plugin. We will learn how to add and use Realm in the project. The gradle plugin dependency should be added to the project scope, which is a project level gradle dependency. Add the following classpath and allow it to sync to your project through the internet:

```
classpath "io.realm:realm-gradle-plugin:3.0.0"
```

Add this dependency to the gradle project level file located as shown in the following screenshot. Add `classpath` to the dependency section:

After adding the class path dependency, the complete gradle file will look as follows:

```
// Top-level build file where you can add configuration options common to
all sub-projects/modules.

buildscript {
    repositories {
        jcenter()
    }
    dependencies {
        classpath 'com.android.tools.build:gradle:2.2.3'

        //Realm plugin
        classpath "io.realm:realm-gradle-plugin:3.0.0"
        // NOTE: Do not place your application dependencies here; they
        belong
        // in the individual module build.gradle files
    }
}

allprojects {
    repositories {
        jcenter()
```

```
    }
}

task clean(type: Delete) {
    delete rootProject.buildDir
}
```

Next, we will apply the Realm plugin to our project to use all the Realm features as shown in the following screenshot:

```
apply plugin: 'realm-android'
```

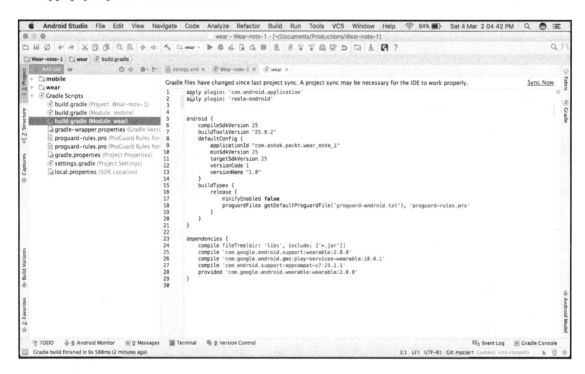

A sample of gradle modified files have been open sourced by the Realm community:

```
https://github.com/realm/realm-java/blob/master/examples/build.g
radle
https://github.com/realm/realm-java/blob/master/examples/introEx
ample/build.gradle
```

Now, we are all set to go for replacing our `SharedPreferences` with Realm. Let's get started.

Open Android Studio and, in the model package, we have defined our `Note.java` POJO class with primitive data string notes and string ID. We have the getters and setters for these variables. Additional changes are as follows:

Extend the POJO model to `RealmObject` and create an empty constructor:

```
public class Note extends RealmObject {

    private String notes = "";
    private String id = "";

    //Empty constructor
    public Note() {

    }

    public Note(String id, String notes) {
        this.id = id;
        this.notes = notes;
    }

    public String getNotes() {
        return notes;
    }

    public void setNotes(String notes) {
        this.notes = notes;
    }

    public String getId() {
        return id;
    }

    public void setId(String id) {
        this.id = id;
    }

}
```

In `MainActivity`, we will instantiate the `Realm` class globally and we will initiate in the `onCreate` method as follows:

```
//MainActivity scope
//Realm Upgrade
private Realm realm;

@Override
```

```
protected void onCreate(Bundle savedInstanceState) {
    super.onCreate(savedInstanceState);
    ....
    //Realm init
    Realm.init(this);
    realm = Realm.getDefaultInstance();
}
```

In the `updateAdapter()` method, we will have to add the read query from Realm, which looks as follows:

```
RealmResults<Note> results = realm.where(Note.class).findAll();
```

The complete method will look as follows:

```
private void updateAdapter() {
    RealmResults<Note> results = realm.where(Note.class).findAll();
    myDataSet.clear();
    myDataSet.addAll(SharedPreferencesUtils.getAllNotes(this));
    mAdapter.setListNote(myDataSet);
    mAdapter.notifyDataSetChanged();
}
```

Realm database provides numerous queries, which maps one to one and one to many relationships for stored data and, for this project, the preceding query does all the magic needed.

In the `createNote()` method, we will change the code to work with Realm instead of `SharedPreference`, as follows:

```
private Note createNote(String id, String note) {
    if (id == null) {
        id = String.valueOf(System.currentTimeMillis());
        realm.beginTransaction();
        Note notes = realm.createObject(Note.class);
        notes.setId(id);
        notes.setNotes(note);
        realm.commitTransaction();
    }
    return new Note(id, note);
}
```

For deleting the records, let's create a new method and call it `deleteRecord()`. In this method, we will pass the ID of the note and delete that note from the Realm:

```
public void deleteRecord(String id){
    RealmResults<Note> results = realm.where(Note.class).equalTo("id",
    id).findAll();

    realm.beginTransaction();

    results.deleteAllFromRealm();

    realm.commitTransaction();
}
```

Now, let's call the delete record method in the `updateData()` method as follows:

```
private void updateData(Note note, int action) {
    if (action == Constants.ACTION_ADD) {
        ConfirmationUtils.showMessage(getString(R.string.note_saved),
        this);
    } else if (action == Constants.ACTION_DELETE) {
        deleteRecord(note.getId());
        ConfirmationUtils.showMessage(getString(R.string.note_removed),
        this);
    }
    updateAdapter();
}
```

The complete `MainActivity` code looks as follows:

```
public class MainActivity extends WearableActivity implements
RecyclerViewAdapter.ItemSelectedListener {

    private static final String TAG = "MainActivity";
    private static final int REQUEST_CODE = 1001;
    private RecyclerViewAdapter mAdapter;
    private List<Note> myDataSet = new ArrayList<>();

    //Realm Upgrade
    private Realm realm;

    @Override
    protected void onCreate(Bundle savedInstanceState) {
        super.onCreate(savedInstanceState);
        setContentView(R.layout.activity_main);
        configLayoutComponents();

        Realm.init(this);
```

```
        realm = Realm.getDefaultInstance();

    }


    private void configLayoutComponents() {
        WearableRecyclerView recyclerView = (WearableRecyclerView)
        findViewById(R.id.wearable_recycler_view);
        recyclerView.setHasFixedSize(true);
        LinearLayoutManager mLayoutManager = new
        LinearLayoutManager(this);
        recyclerView.setLayoutManager(mLayoutManager);


        mAdapter = new RecyclerViewAdapter();
        mAdapter.setListNote(myDataSet);
        mAdapter.setListener(this);
        recyclerView.setAdapter(mAdapter);

        EditText editText = (EditText) findViewById(R.id.edit_text);

        editText.setOnEditorActionListener(new
        TextView.OnEditorActionListener() {
            @Override
            public boolean onEditorAction(TextView textView, int
            action, KeyEvent keyEvent) {
                if (action == EditorInfo.IME_ACTION_SEND) {
                    String text = textView.getText().toString();
                    if (!TextUtils.isEmpty(text)) {
                        Note note = createNote(null, text);
                        SharedPreferencesUtils.saveNote(note,
                        textView.getContext());
                        updateData(note, Constants.ACTION_ADD);
                        textView.setText("");
                        return true;
                    }
                }
                return false;
            }
        });
    }

    private void updateAdapter() {
        RealmResults<Note> results = realm.where(Note.class).findAll();
        myDataSet.clear();
        myDataSet.addAll(results);
        mAdapter.setListNote(myDataSet);
        mAdapter.notifyDataSetChanged();
```

```
    }

@Override
protected void onResume() {
    super.onResume();
    updateAdapter();
}

@Override
public void onItemSelected(int position) {
    Intent intent = new Intent(getApplicationContext(),
    DeleteActivity.class);
    intent.putExtra(Constants.ITEM_POSITION, position);
    startActivityForResult(intent, REQUEST_CODE);
}

@Override
protected void onActivityResult(int requestCode, int resultCode,
Intent data) {
    if (data != null && requestCode == REQUEST_CODE && resultCode
    == RESULT_OK) {
        if (data.hasExtra(Constants.ITEM_POSITION)) {
            int position =
            data.getIntExtra(Constants.ITEM_POSITION, -1);
            if (position > -1) {
                Note note = myDataSet.get(position);
                updateData(note, Constants.ACTION_DELETE);
            }
        }
    }
}


private void prepareUpdate(String id, String title, int action) {
    if (!(TextUtils.isEmpty(id) && TextUtils.isEmpty(title))) {
        Note note = createNote(id, title);
        updateData(note, action);
    }
}

private void updateData(Note note, int action) {
    if (action == Constants.ACTION_ADD) {
        ConfirmationUtils.showMessage
        (getString(R.string.note_saved), this);

    } else if (action == Constants.ACTION_DELETE) {
        deleteRecord(note.getId());
        ConfirmationUtils.showMessage(getString
```

```
            (R.string.note_removed), this);
        }
        updateAdapter();
    }

    /**
     * Notifica a Data Layer API que os dados foram modificados.
     */


    private Note createNote(String id, String note) {
        if (id == null) {
            id = String.valueOf(System.currentTimeMillis());
            realm.beginTransaction();
            Note notes = realm.createObject(Note.class);
            notes.setId(id);
            notes.setNotes(note);
            realm.commitTransaction();
        }
        return new Note(id, note);
    }

    public void deleteRecord(String id){
        RealmResults<Note> results = realm.where(Note.class)
        .equalTo("id", id).findAll();

        realm.beginTransaction();

        results.deleteAllFromRealm();

        realm.commitTransaction();
    }

    @Override
    protected void onDestroy() {
        realm.close();
        super.onDestroy();
    }
}
```

Now, functionally, we have Realmdb integrated to our Wear-note-2 app. Let's compile and see the result.

The home screen of the Wear note-taking application looks as follows:

The IMF Screen that the Wear operating system handles to get the input from the user:

The 'Saved animation' through default `confirmationActivity` in the Wear support library:

Home screen with a saved note:

Now, we have replaced `Sharedpreference` with the best database of our time for Android.

Working on the Wear User Interface

In the Wear-note application, we are using a white color background and a black color text with the default Roboto font. When preparing a good Wear application design, Google recommends using a dark color for the Wear application for the best battery efficiency. Light color schemes used in typical material designed mobile applications are not energy efficient in Wear devices. Light colors are less energy efficient in OLED displays.

Light colors need to light up the pixels with brighter intensity. White colors need to light up the RGB diodes in the pixels at 100%; the more white and light color in the application, the lesser battery efficient the application will be.

Light colors are disruptive in dark light or when using your Wear device at night. This scenario would probably not happen with dark colors. Unlike light colors, dark colors make the screen less bright when they are active and saves battery in OLED displays.

Let's get started working on the Wear-note-2 user interface

Let's change the application theme for adapting the standard design guidelines.

In the `activity_main.xml` file, we will edit the parent container's background, which is the `BoxInsetLayout` background `android:background="#01579B"` to cobalt blue.

Add a new `color.xml` under the `values` directory and add the following code:

```
//Add this color value in the color.xml in res directory
<color name="cobalt_blue">#01579B</color>
```

After adding the color we can use the color in our production application as shown below:

```
<android.support.wearable.view.BoxInsetLayout
    xmlns:android="http://schemas.android.com/apk/res/android"
    xmlns:tools="http://schemas.android.com/tools"
    xmlns:app="http://schemas.android.com/apk/res-auto"
    android:layout_width="match_parent"
    android:layout_height="match_parent"
    android:id="@+id/container"
    tools:context="com.ashok.packt.wear_note_1.activity.MainActivity"
    tools:deviceIds="wear"
    android:background="@color/cobalt_blue"
    app:layout_box="all"
    android:padding="5dp">
```

Further, let's change the color of the `EditText` hint color, as follows:

```
<EditText
    android:id="@+id/edit_text"
    android:layout_width="match_parent"
    android:layout_height="50dp"
    android:layout_gravity="center"
    android:gravity="center"
    android:hint="@string/add_a_note"
    android:imeOptions="actionSend"
    android:inputType="textCapSentences|textAutoCorrect"
    android:textColor="@color/white"
    android:textColorHint="@color/white"
    android:layout_alignParentTop="true"
    android:layout_alignParentStart="true" />
```

In the `each_item.xml` layout, modify the XML code as follows:

```
<?xml version="1.0" encoding="utf-8"?>
<LinearLayout
    xmlns:android="http://schemas.android.com/apk/res/android"
    xmlns:tools="http://schemas.android.com/tools"
    android:layout_width="match_parent"
    android:layout_height="match_parent"
    android:gravity="center"
    android:layout_gravity="center"
    android:clickable="true"
    android:background="?android:attr/selectableItemBackground"
```

```
android:orientation="vertical">

<TextView
    android:id="@+id/note"
    android:layout_width="wrap_content"
    android:layout_height="wrap_content"
    android:gravity="center"
    android:layout_gravity="center"
    android:textColor="@color/white"
    tools:text="note"/>

</LinearLayout>
```

Now, do the same changes in the `activity_delete.xml` layout's container, change its background color and change the color of `TextView`. Changing the color of `DelayedConfirmationView` can be done using the `xmlns` properties, as highlighted in the following code:

```
<android.support.wearable.view.DelayedConfirmationView
    android:id="@+id/delayed_confirmation"
    android:layout_width="wrap_content"
    android:layout_height="wrap_content"
    android:src="@android:drawable/ic_delete"
    app:circle_border_color="@color/white"
    app:circle_color="@color/white"
    app:circle_border_width="8dp"
    app:circle_radius="30dp"/>
```

Developers need not worry about changing the color of animations in `DelayedConfirmationView`, Wear 2.0 automatically adapts to the theme of `DelayedConfirmationView` and changes the primary color schemes for creating a unified experience.

All these changes reflect in the application as follows:

IMF screen for fetching input from the user:

After saving the data using Realm:

Deleting the note from the database:

Better fonts for better reading

In the world of digital design, making your application's visuals easy on users' eyes is important. The Lora font from Google collections has well-balanced contemporary serifs with roots in calligraphy. It is a text typeface with moderate contrast well suited for body text. A paragraph set in Lora will be memorable because of its brushed curves in contrast with driving serifs. The overall typographic voice of Lora perfectly conveys the mood of a modern-day story, or an art essay. Technically, Lora is optimized for screen appearance.

To add the custom fonts to an Android project, we need to create the `assets` folder in root. Check the following screenshot:

We can directly add the `.ttf` file assets directory or we can create another directory and the fonts. You can download the Lora font via this URL: `https://fonts.google.com/downloa d?family=Lora`.

After adding the font files to the asset folder, we need to create custom `Textview` and `EditText`.

In the `utils` package, let's create the class called `LoraWearTextView` and `LoraWearEditTextView`, as follows:

Now, extend `LoraWearTextView` to the `TextView` class and `LoraWearEditTextView` to `EditText`. Implement the constructor methods in both the classes. Create a new method and call it `init()`. Inside the `init` method, instantiate the `Typeface` class and, using the `createFromAsset` method, we can set our custom font:

```
public void init() {
    Typeface tf = Typeface.createFromAsset(getContext().getAssets(),
    "fonts/Lora.ttf");
    setTypeface(tf ,1);

}
```

The preceding `init` method is the same in both the classes. Call the `init` method in all the different parameterised constructors of the two new custom classes.

The complete class looks as follows: `LoraWearTextView.java`

```
public class LoraWearTextView extends TextView {
    public LoraWearTextView(Context context) {
        super(context);
        init();
    }

    public LoraWearTextView(Context context, AttributeSet attrs) {
        super(context, attrs);
        init();
    }
```

```java
    public LoraWearTextView(Context context, AttributeSet attrs, int
    defStyleAttr) {
        super(context, attrs, defStyleAttr);
        init();
    }

    public LoraWearTextView(Context context, AttributeSet attrs, int
    defStyleAttr, int defStyleRes) {
        super(context, attrs, defStyleAttr, defStyleRes);
        init();
    }

    public void init() {
        Typeface tf = Typeface.createFromAsset(getContext()
        .getAssets(), "fonts/Lora.ttf");
        setTypeface(tf ,1);

    }
}
```

The complete class looks as follows: `LoraWearEditTextView.java`

```java
public class LoraWearEditTextView extends EditText {


public LoraWearEditTextView(Context context) {
 super(context);
 init();
}

public LoraWearEditTextView(Context context, AttributeSet attrs) {
 super(context, attrs);
 init();
}

public LoraWearEditTextView(Context context, AttributeSet attrs, int
defStyleAttr) {
 super(context, attrs, defStyleAttr);
 init();
}

public LoraWearEditTextView(Context context, AttributeSet attrs, int
defStyleAttr, int defStyleRes) {
 super(context, attrs, defStyleAttr, defStyleRes);
 init();
}
```

```
public void init() {
  Typeface tf = Typeface.createFromAsset(getContext().getAssets(),
  "fonts/Lora.ttf");
  setTypeface(tf ,1);
}
```

After adding these two classes in our UI layouts, we can replace the actual `textview` and `edittext`:

```
<com.ashok.packt.wear_note_1.utils.LoraWearEditTextView
    android:id="@+id/edit_text"
    android:layout_width="match_parent"
    android:layout_height="50dp"
    android:layout_gravity="center"
    android:gravity="center"
    android:hint="@string/add_a_note"
    android:imeOptions="actionSend"
    android:inputType="textCapSentences|textAutoCorrect"
    android:textColor="@color/white"
    android:textColorHint="@color/white"
    android:layout_alignParentTop="true"
    android:layout_alignParentStart="true" />

<com.ashok.packt.wear_note_1.utils.LoraWearTextView
    android:id="@+id/note"
    android:layout_width="wrap_content"
    android:layout_height="wrap_content"
    android:gravity="center"
    android:layout_gravity="center"
    android:textColor="@color/white"
    tools:text="note"/>
```

After replacing `textview` and `edittext`, compile the program and let's see the result:

Remove Note animation through `confirmationViewActivity`:

In this chapter, we have explored how to integrate the Realm popular database, understood the design philosophy for wear devices, and created our custom view components for better application user experience.

Writing custom layouts for better UX

Android offers great ways to customize the layouts that we use as containers. We can have compound views and customize the layout for our own purposes. Wearables use the same layout techniques as handheld Android devices, but need to be designed with specific constraints and configuration. It's not a good idea to port a functionality from a handheld android device component to Wear app designs. For designing a great Wear app, follow the Google design guidelines at `https://developer.android.com/design/wear/index.html`. Let's create custom views and use them in the Wear note application.

Let's implement animations in our layouts with the way we are loading the items. We shall have a simple slide animation and we will do that with the custom layout we write. Let's create a class file called `AnimatedLinearLayout` and extend it to `LinearLayout` as follows:

 Compound views are the combination of two or more views bundled as one component, for instance, the Checkable Relative layout. When a user clicks on the layout, it will highlight the layout similar to the checkbox.

```
public class AnimatedLinearLayout extends LinearLayout {

...

}
```

Now, we need the `Animation` class from `View`. Apart from the `Animation` class, declare the `View` instance for a `currentChild` view. Since we are writing a layout and it can hold the child's hierarchy, we need a view instance for reference:

```
Animation animation;
View currentChild;
```

When we extend the class to `LinearLayout`, we get a couple of constructor `callbacks`, which we have to implement:

```
public AnimatedLinearLayout(Context context) {
    super(context);
}

public AnimatedLinearLayout(Context context, AttributeSet attrs) {
    super(context, attrs);
}

@Override
public void onWindowFocusChanged(boolean hasWindowFocus) {
    super.onWindowFocusChanged(hasWindowFocus);
    ...
}
```

In the `onWindowFocusChanged()` method, we can write our logic for the custom animation. Here, the book introduces `SlideDown`, `SlideDownMore`, `RotateClockWise`, `RotateAntiClockWise`, and `ZoomInAndRotateClockWise`. Now, to implement this, we need to check whether the view is inflated and has got the window to be displayed, and how many `childviews` the layout has:

```
if (hasWindowFocus) {
    for (int index = 0; index < getChildCount(); index++) {
        View child = getChildAt(index);
        currentChild=child;
}
```

Check whether the child is an instance of Viewgroup; if the view is developed completely in any other alienated way, then this custom layout will not be able to help that view to animate. Using the childviews tag property, we can assign a string association for animations, as follows:

```
if(!(child instanceof ViewGroup)) {
    switch (child.getTag().toString()) {
    case SLIDE_DOWN:
      // write logic to slide down animation
```

For the slide down animation, using the animation instance we created globally, we have to set the interpolator and decelerate the interpolation with the childviews height by the numeric value 2. Check the following code:

```
case SLIDE_DOWN:
    animation = new TranslateAnimation(0, 0, -
    ((child.getMeasuredHeight()/2) * (index + 1)), 0);
    animation.setInterpolator(new DecelerateInterpolator());
    animation.setFillAfter(true);
    animation.setDuration(1000);
    child.post(new AnimationRunnable(animation,child));
    //child.startAnimation(animation);
    break;
```

Similarly, we will complete other cases. Let's work on completing the onWindowFocusChanged() method:

```
@Override
public void onWindowFocusChanged(boolean hasWindowFocus) {
    super.onWindowFocusChanged(hasWindowFocus);
    final String SLIDE_DOWN = "SlideDown";
    final String SLIDE_DOWN_MORE = "SlideDownMore";
    final String ROTATE_CLOCKWISE = "RotateClockWise";
    final String ROTATE_ANTI_CLOCKWISE = "RotateAntiClockWise";
    final String ZOOMIN_AND_ROTATE_CLOCKWISE =
    "ZoomInAndRotateClockWise";
    if (hasWindowFocus) {
        for (int index = 0; index < getChildCount(); index++) {
            View child = getChildAt(index);
            currentChild=child;
            if(!(child instanceof ViewGroup)) {
                switch (child.getTag().toString()) {
                    case SLIDE_DOWN:
                        animation = new TranslateAnimation(0, 0, -
                        ((child.getMeasuredHeight()/2) *
                        (index + 1)), 0);
                        animation.setInterpolator(new
                        DecelerateInterpolator());
                        animation.setFillAfter(true);
                        animation.setDuration(1000);
                        child.post(new
                        AnimationRunnable(animation,child));
                        //child.startAnimation(animation);
                        break;
                    case SLIDE_DOWN_MORE:
                        animation = new TranslateAnimation(0, 0, -
                        (child.getMeasuredHeight() * (index + 25)), 0);
                        animation.setInterpolator(new
```

```
        DecelerateInterpolator());
        animation.setFillAfter(true);
        animation.setDuration(1000);
        child.post(new
        AnimationRunnable(animation,child));
        //child.startAnimation(animation);
        break;
    case ROTATE_CLOCKWISE:
        animation = new RotateAnimation(0, 360,
        child.getMeasuredWidth() / 2,
        child.getMeasuredHeight() / 2);
        animation.setInterpolator(new
        DecelerateInterpolator());
        animation.setFillAfter(true);
        animation.setDuration(1000);
        child.post(new
        AnimationRunnable(animation,child));
        //child.startAnimation(animation);
        break;
    case ROTATE_ANTI_CLOCKWISE:
        animation = new RotateAnimation(0, -360,
        child.getMeasuredWidth() / 2,
        child.getMeasuredHeight() / 2);
        animation.setInterpolator(new
        DecelerateInterpolator());
        animation.setFillAfter(true);
        animation.setDuration(1000);
        child.post(new
        AnimationRunnable(animation,child));
        //child.startAnimation(animation);
        break;
    case ZOOMIN_AND_ROTATE_CLOCKWISE:
        AnimationSet animationSet = new
        AnimationSet(true);
        animationSet.setInterpolator(new
        DecelerateInterpolator());
        animation = new ScaleAnimation(0, 1, 0, 1,
        child.getMeasuredWidth() / 2,
        child.getMeasuredHeight() / 2);
        animation.setStartOffset(0);
        animation.setFillAfter(true);
        animation.setDuration(1000);
        animationSet.addAnimation(animation);
        animation = new RotateAnimation(0, 360,
        child.getMeasuredWidth() / 2,
        child.getMeasuredHeight() / 2);
        animation.setStartOffset(0);
        animation.setFillAfter(true);
```

```
                                    animation.setDuration(1000);
                                    animationSet.addAnimation(animation);
                                    child.post(new
                                    AnimationSetRunnable(animationSet,child));
                                    //child.startAnimation(animationSet);
                                    break;
                            }
                        }
                    }
                }
            }
```

Now, we need to create an `AnimationRunnable` class, which implements the `Runnable` interface to start the animation:

```
    private class AnimationRunnable implements Runnable{
        private Animation animation;
        private View child;
        AnimationRunnable(Animation animation, View child) {
            this.animation=animation;
            this.child=child;
        }

        @Override
        public void run() {
            child.startAnimation(animation);
        }
    }
```

We will implement another `AnimationSetRunnable` class to the runnable interface for setting the animation:

```
    private class AnimationSetRunnable implements Runnable{
        private AnimationSet animation;
        private View child;
        AnimationSetRunnable(AnimationSet animation, View child) {
            this.animation=animation;
            this.child=child;
        }

        @Override
        public void run() {
            child.startAnimation(animation);
        }
    }
```

Now, we have completed our own custom layout that has a couple of animation methods to all our view childs in the layout. The complete class code for this custom layout looks as follows:

```
public class AnimatedLinearLayout extends LinearLayout {
    Animation animation;
    View currentChild;

    public AnimatedLinearLayout(Context context) {
        super(context);
    }

    public AnimatedLinearLayout(Context context, AttributeSet attrs) {
        super(context, attrs);
    }

    @Override
    public void onWindowFocusChanged(boolean hasWindowFocus) {
        super.onWindowFocusChanged(hasWindowFocus);
        final String SLIDE_DOWN = "SlideDown";
        final String SLIDE_DOWN_MORE = "SlideDownMore";
        final String ROTATE_CLOCKWISE = "RotateClockWise";
        final String ROTATE_ANTI_CLOCKWISE = "RotateAntiClockWise";
        final String ZOOMIN_AND_ROTATE_CLOCKWISE =
        "ZoomInAndRotateClockWise";
        if (hasWindowFocus) {
            for (int index = 0; index < getChildCount(); index++) {
                View child = getChildAt(index);
                currentChild=child;
                if(!(child instanceof ViewGroup)) {
                    switch (child.getTag().toString()) {
                        case SLIDE_DOWN:
                            animation = new TranslateAnimation(0, 0, -
                            ((child.getMeasuredHeight()/2) *
                             (index + 1)), 0);
                            animation.setInterpolator(new
                            DecelerateInterpolator());
                            animation.setFillAfter(true);
                            animation.setDuration(1000);
                            child.post(new
                            AnimationRunnable(animation,child));
                            //child.startAnimation(animation);
                            break;
                        case SLIDE_DOWN_MORE:
                            animation = new TranslateAnimation(0, 0, -
                            (child.getMeasuredHeight() *
                            (index + 25)), 0);
                            animation.setInterpolator(new
```

```
                DecelerateInterpolator());
                animation.setFillAfter(true);
                animation.setDuration(1000);
                child.post(new
                AnimationRunnable(animation,child));
                //child.startAnimation(animation);
                break;
        case ROTATE_CLOCKWISE:
                animation = new RotateAnimation(0, 360,
                child.getMeasuredWidth() / 2,
                child.getMeasuredHeight() / 2);
                animation.setInterpolator(new
                DecelerateInterpolator());
                animation.setFillAfter(true);
                animation.setDuration(1000);
                child.post(new
                AnimationRunnable(animation,child));
                //child.startAnimation(animation);
                break;
        case ROTATE_ANTI_CLOCKWISE:
                animation = new RotateAnimation(0, -360,
                child.getMeasuredWidth() / 2,
                child.getMeasuredHeight() / 2);
                animation.setInterpolator(new
                DecelerateInterpolator());
                animation.setFillAfter(true);
                animation.setDuration(1000);
                child.post(new
                AnimationRunnable(animation,child));
                //child.startAnimation(animation);
                break;
        case ZOOMIN_AND_ROTATE_CLOCKWISE:
                AnimationSet animationSet = new
                AnimationSet(true);
                animationSet.setInterpolator(new
                DecelerateInterpolator());
                animation = new ScaleAnimation(0, 1, 0, 1,
                child.getMeasuredWidth() / 2,
                child.getMeasuredHeight() / 2);
                animation.setStartOffset(0);
                animation.setFillAfter(true);
                animation.setDuration(1000);
                animationSet.addAnimation(animation);
                animation = new RotateAnimation(0, 360,
                child.getMeasuredWidth() / 2,
                child.getMeasuredHeight() / 2);
                animation.setStartOffset(0);
                animation.setFillAfter(true);
```

```
                        animation.setDuration(1000);
                        animationSet.addAnimation(animation);
                        child.post(new
                        AnimationSetRunnable(animationSet,child));
                        //child.startAnimation(animationSet);
                        break;
                    }
                }
            }
        }
    }

    private class AnimationRunnable implements Runnable{
        private Animation animation;
        private View child;
        AnimationRunnable(Animation animation, View child) {
            this.animation=animation;
            this.child=child;
        }

        @Override
        public void run() {
            child.startAnimation(animation);
        }
    }
    private class AnimationSetRunnable implements Runnable{
        private AnimationSet animation;
        private View child;
        AnimationSetRunnable(AnimationSet animation, View child) {
            this.animation=animation;
            this.child=child;
        }

        @Override
        public void run() {
            child.startAnimation(animation);
        }
    }
}
```

Now, after completely writing the class, it is time to use it in our project. We can use this classname as an XML tag and use it in the each_item.xml layout, which is the row item layout of recyclerview:

```
<com.ashok.packt.wear_note_1.utils.AnimatedLinearLayout
    xmlns:android="http://schemas.android.com/apk/res/android"
    xmlns:tools="http://schemas.android.com/tools"
    android:orientation="vertical">
```

```
</com.ashok.packt.wear_note_1.utils.AnimatedLinearLayout>
```

Replace the layout code with the new `AnimatedLinearLayout`. We need to pass the tag in `childviews` for getting animated. The following code explains this in detail:

```xml
<?xml version="1.0" encoding="utf-8"?>
<com.ashok.packt.wear_note_1.utils.AnimatedLinearLayout
xmlns:android="http://schemas.android.com/apk/res/android"
    xmlns:tools="http://schemas.android.com/tools"
    android:layout_width="match_parent"
    android:layout_height="match_parent"
    android:layout_gravity="center"
    android:background="?android:attr/selectableItemBackground"
    android:clickable="true"
    android:gravity="center"
    android:orientation="vertical">

    <com.ashok.packt.wear_note_1.utils.AnimatedLinearLayout
        android:layout_width="match_parent"
        android:layout_height="wrap_content"
        android:layout_marginTop="10dp"
        android:gravity="center"
        android:orientation="horizontal">

        <com.ashok.packt.wear_note_1.utils.LoraWearTextView
            android:id="@+id/note"
            android:layout_width="wrap_content"
            android:layout_height="wrap_content"
            android:layout_gravity="center"
            android:gravity="center"
            android:tag="ZoomInAndRotateClockWise"
            android:textColor="@color/white"
            tools:text="note" />
    </com.ashok.packt.wear_note_1.utils.AnimatedLinearLayout>

</com.ashok.packt.wear_note_1.utils.AnimatedLinearLayout>
```

This layout will draw all the views with small animations and display the list item.

The `ZoomInAndRotateClockWise` animation and try changing the string exactly the same string given in the Custom class:

Summary

In this chapter, we have understood the importance of dark themes on Wear devices. We changed the custom `TextView` with font assets. We have seen how to write custom layouts and defined a few animations in it. In later projects, we will explore more of the design guidelines and components introduced exclusively in Wear 2.0 `Wearable action drawer` and `wearable navigation drawer`.

4
Measure Your Wellness - Sensors

We are living in the realm of technology! It is definitely not the highlight here. We are also living in the realm of intricate lifestyles that are driving everyone's health into some sort of illness. Our existence leads us back to the roots of the ocean; we all know that we are beings who have evolved from water. If we trace back, we clearly understand that our body composition is made of 60 percent water and the rest is muscled water. When we talk about taking care of our health, we miss simple things, such as drinking sufficient water. Adequate, regular water consumption will ensure a great metabolism and healthy, functional organs. Then new millennium's advancements in technologies are an expression of how one can make use of technology for doing the right things.

Android Wear integrates numerous sensors, which can be used to help Android Wear users measure their heart rate, step counts, and much more. Having said that, how about writing an application that reminds us to drink water every 30 minutes, measures our heart rate and step count, and offers a few health tips?

This chapter will enable you to understand how to enlist all the available sensors in the Wear device and use them for measuring step counts, heart rate, and more.

In this chapter, we will explore the following:

- Enlisting the available sensors in Wear
- The accuracy of the sensors and battery consumption
- Wear 2.0 doze mode

- Writing the app with initial logic
- Getting started on the material design for Wear
- Creating a user interface for the app

Conceptualizing the application

The Wear app, which we will build, will have three main interfaces for starting the water reminder, heart rate analysis, and step counter.

The following is the water reminder screen for starting the reminder service. Using the navigation facility in the navigation drawer, we can swap between other screens:

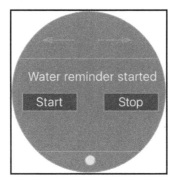

Enlisting the available sensors in Wear

Knowing the list of sensors available in the Wear device you are working on is actually good ensuring you do for not have false results and unnecessary waiting. Few android Wear devices don't have the heart rate sensor. Writing an app for the heart rate sensor, in such cases, proves nothing. Here's how we can get the list of available sensors:

```
public class SensorActivity extends Activity implements SensorEventListener
{
  private SensorManager mSensorManager;
  private Sensor mLight;

  @Override
  public final void onCreate(Bundle savedInstanceState) {
    super.onCreate(savedInstanceState);
    setContentView(R.layout.main);
```

```
mSensorManager = (SensorManager)
getSystemService(Context.SENSOR_SERVICE);
mLight = mSensorManager.getDefaultSensor(Sensor.TYPE_LIGHT);

mSensorManager = (SensorManager)
getSystemService(Context.SENSOR_SERVICE);
final List<Sensor> deviceSensors =
mSensorManager.getSensorList(Sensor.TYPE_ALL);

for(Sensor type : deviceSensors){
    Log.e("sensors",type.getStringType());
}
}

@Override
public final void onAccuracyChanged(Sensor sensor, int accuracy){
  // Do something here if sensor accuracy changes.
}

@Override
public final void onSensorChanged(SensorEvent event) {
}

@Override
protected void onResume() {
  super.onResume();

}

@Override
protected void onPause() {
  super.onPause();
  mSensorManager.unregisterListener(this);
}
}
```

We will see all the available sensors in the Wear device in the console, as follows:

```
E/sensors: android.sensor.accelerometer
E/sensors: android.sensor.magnetic_field
E/sensors: android.sensor.orientation
E/sensors: android.sensor.temperature
E/sensors: android.sensor.proximity
E/sensors: android.sensor.light
E/sensors: android.sensor.pressure
E/sensors: android.sensor.relative_humidity
E/sensors: android.sensor.geomagnetic_rotation_vector
```

Accuracy of sensors

All the sensors packed into the Wear devices exhibit the best possible precision and accuracy. When we are writing an application for the health field, considering the accuracy of the sensor is very important. The sensors used in Wear devices are also dependent on manufacturers since Android Wear has produced nearly 40 and more different brands as per the Google IO 2017. When we are writing the application from the groundup without knowledge of what Wear offers, we surely will experience many challenges. Most of the Wear devices have step detector and step counter sensors. Perhaps, we need not worry about writing an accelerometer program that takes the three-dimensional motion of the device and predicts the step, which will be inaccurate. Linear acceleration predicted steps can lead to false results most of the time. For saving the battery and making a meaningful decision in certain use cases, sensors play a vital role. Motion sensors help in saving the Wear device's battery by turning the ambient mode on when the wrist is not being looked at. Light sensors will allow devices to increase or decrease the screen's brightness according to the external light influence. Heart rate sensors are common in Wearable devices. Essentially, a heart rate sensor is an optical sensor. It will not be as accurate as an electrocardiogram or electrocardiograph (EKG), but these sensors will be close to accurate when you are using them during periods of rest. Using the same sensors, sleep monitoring and activity tracking, such as calorie burning, can be predicted. These sensors are getting better and will be extremely accurate in the future, but, for the quality of the hardware, what is packed in Wear will receive software fixes for hardware issues and software level controls.

The common Wear integrated sensors are as follows:

- Accelerometer
- Magnetometer
- Ambient sensor
- GPS
- Heart rate sensor
- Oxygen saturation sensor

In the future, Wear devices will pack in more sensors with more accuracy and they will target the health area through these sensors.

Battery consumption

The battery is vital. The programs we write should target saving the battery. The Wear 1.x design guidelines and standards didn't have material design and it had a huge impact on battery consumption. After the material design was introduced to Wear 2.0, it became clear that a dark theme saves more battery. The Android Wear team is continuously working on improving the battery life. Google has introduced a set of best practices to improve the battery life of Wear devices. The display in Wear devices consumes a lot of battery. There are different types of display and they draw different quantities of power when the display is in the on mode, in the always on mode, and in the interactive mode. The interactive mode draws the maximum power, so we need to be concerned about when to write the interactive mode. Of course, battery consumption is necessary for most use cases. When we really look into a particular use case, there will be numerous places where we can stop the max power being drawn from the Wear battery. We should always double-check in the production app that we are releasing all the sensors and other hardware that will consumed battery. Android offers the WAKE_LOCK mechanism, which enables developers to identify which app is using which hardware, and helps to release them when the application goes into the background. Writing services are necessary for running a long process, but if we utilize background services for hardware acquisition, they will always be consuming the battery, and writing a kill service in such cases is very important.

Android Studio offers an open source tool called Battery Historian https://developer.android.com/studio/profile/battery-historian.html for analyzing the battery stats for apps. Battery historian converts battery data into HTML visual representations that can be viewed on a browser. Before trying to save the battery, we need to know which process draws how much current from the battery. Later, using the battery historian data, we will determine which process is consuming more data and we will be able to work on battery optimization.

Doze mode

Battery optimization is always a challenging task for developers. Google has initiated many in-house projects, such as Project Volta, which insists on an animation rate of 60 frames per second. Manufacturers have started a trend of releasing phones and tablets packed with the highest battery capacity. The challenge here was that there were too many new hardware components introduced and the devices were long lasting with a few more hours. Instead of packing an enormous battery, Google optimized the battery consumption process from API level 23 to accommodate few more hours of usage when the applications are adapted to use the doze mode.

When the device is not being used for a long time or when the device is in an idle state for a long time, the device enters into doze mode. When the system tries to access network and CPU-intensive services, the doze mode prevents this from happening. Doze mode exits automatically after a periodic time, accesses the network, and syncs all the jobs and alarms. The following restrictions apply to your app when it is in the mode:

- Network operations are killed
- The system ignores wake locks
- Standard alarm managers need to wait for the doze mode to be exited
- To fire an alarm when it is in doze mode, it should use `setAndAllowWhileIdle()`
- The device will not perform Wi-Fi scans
- The sync adapters are restricted
- The job schedulers are restricted

Adapting the doze mode to your applications can affect them differently depending on the capabilities they offer. Almost all the apps work normally when they are in doze mode. In some use cases, we have to modify the network request jobs, alarm, and syncs, and the app should manage all the activities efficiently during the doze mode exit. Doze mode is likely to influence the activities that the `AlarmManager` alerts for and timers manage, considering that alarms in Android 5.1 (API level 22) or lower do not fire when the system is in doze mode. To help schedule alarms, Android 6.0 (API level 23) introduces two new `AlarmManager` methods: `setAndAllowWhileIdle()` and `setExactAndAllowWhileIdle()`. With these methods, we can fire an alarm in doze mode.

Creating a project

Now, let's fire up the Android Studio and build a Wear app that helps us to do all this.

1. Create an Android project and call it `Upbeat`. Since it is a health-related application, the name upbeat makes a lot of sense:

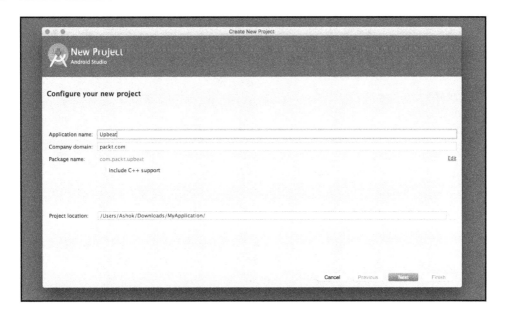

2. Choose a **Phone** and **Wear** module in the **Target Android Devices** screen:

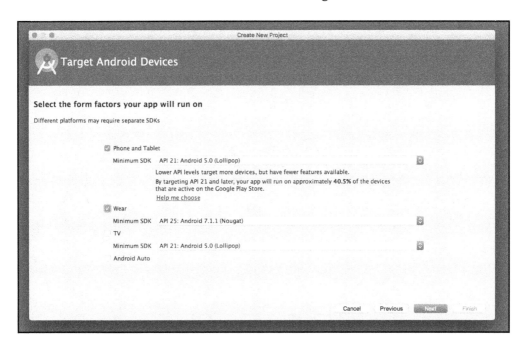

4. Now choose the **Empty Activity** for the mobile template and the **Always On Wear Activity** for Wear. After your project is successfully created, go to the Wear module and add the color scheme file to start driving the development of the material design standard.

5. Create a `colors.xml` file in the `res/values` directory and add the following color codes:

```xml
<?xml version="1.0" encoding="utf-8"?>
<resources>
    <color name="colorPrimary">#607d8b</color>
    <color name="colorPrimaryDark">#34515e</color>
    <color name="colorAccent">#FFF</color>
</resources>
```

For better readability of code for the project, let's create packages called `fragments`, `services`, and `utils`. In these three packages, we will create all our working code.

6. In this project, we will use `WearableDrawerLayout` and `WearableNavigationDrawer` for navigating between fragments. Let's set up `MainActivity` in the wear module for working with `WearableNavigationDrawer`. In the `activity_main.xml` file, we need to change the root element to `WearableDrawerLayout`, as follows:

```xml
<android.support.wearable.view.drawer.WearableDrawerLayout
    xmlns:android="http://schemas.android.com/apk/res/android"
    xmlns:app="http://schemas.android.com/apk/res-auto"
    xmlns:tools="http://schemas.android.com/tools"
    android:id="@+id/drawer_layout"
    android:layout_width="match_parent"
    android:layout_height="match_parent"
    android:background="@color/colorPrimary"
    tools:context="com.packt.upbeat.MainActivity"
    tools:deviceIds="wear">

    .
    .
    .

</android.support.wearable.view.drawer.WearableDrawerLayout>
```

7. Inside `nestedScrollview` from the support library, we shall add `framelayout` as a container for the fragments to be attached and detached. After that, we can attach another child of the `WearableNavigationDrawer` element. In case we are looking for the Wear app to have an action menu, which we are, we shall add another element called `WearableActionDrawer`. Add the following code within in the `WearableDrawerLayout` scope:

```
<android.support.wearable.view.drawer.WearableDrawerLayout
  ...

<android.support.v4.widget.NestedScrollView
  android:id="@+id/content"
  android:layout_width="match_parent"
  android:layout_height="match_parent"
  android:fillViewport="true">

  <FrameLayout
  android:id="@+id/content_frame"
  android:layout_width="match_parent"
  android:layout_height="match_parent" />

</android.support.v4.widget.NestedScrollView>

<android.support.wearable.view.drawer.WearableNavigationDrawer
  android:id="@+id/top_navigation_drawer"
  android:layout_width="match_parent"
  android:layout_height="match_parent"
  android:background="@color/colorPrimaryDark" />

<android.support.wearable.view.drawer.WearableActionDrawer
  android:id="@+id/bottom_action_drawer"
  android:layout_width="match_parent"
  android:layout_height="match_parent"
  android:background="@color/colorPrimaryDark"
  app:action_menu="@menu/drawer_menu" />

</android.support.wearable.view.drawer.WearableDrawerLayout>
```

Adding a drawer menu

We are going to add the `drawer_menu` menu XML resource in a moment. All visual elements are in place for `MainActivity`. Let's work on the Java side of the story to make the dynamic swipe and fragments switch for the selected navigation drawer item. Before we get started on `MainActivity`, we need to create a POJO class for the drawer item with a constructor:

```
public class DrawerItem {
    private String name;
    private String navigationIcon;

    public DrawerItem(String name, String navigationIcon) {
        this.name = name;
        this.navigationIcon = navigationIcon;
    }

    public String getName() {
        return name;
    }

    public void setName(String name) {
        this.name = name;
    }

    public String getNavigationIcon() {
        return navigationIcon;
    }

    public void setNavigationIcon(String navigationIcon) {
        this.navigationIcon = navigationIcon;
    }
}
```

These getters and setters will help in setting the drawer icon and drawer title, as we have discussed what a POJO is in a note-taking app earlier.

In `MainActivity.java`, let's implement `WearableActionDrawer.OnMenuItemClickListener` and override `onMenuItemClick`, as follows:

```
public class MainActivity extends WearableActivity  implements
        WearableActionDrawer.OnMenuItemClickListener{


    . . . .
```

```
@Override
public boolean onMenuItemClick(MenuItem menuItem) {
    Log.d(TAG, "onMenuItemClick(): " + menuItem);
    final int itemId = menuItem.getItemId();
  }
}
```

Let's initialize all the instances for `WearableDrawerLayout`,
`WearableNavigationDrawer`, and `WearableActionDrawer` with the necessary setup
inside the scope of `MainActivity`:

```
private static final String TAG = "MainActivity";
private WearableDrawerLayout mWearableDrawerLayout;
private WearableNavigationDrawer mWearableNavigationDrawer;
private WearableActionDrawer mWearableActionDrawer;
private ArrayList<DrawerItem> drawer_itemArrayList;
private int mSelectedScreen;
```

In the `onCreate` method, let's map all the visual components through the
`findViewById()` method. Thereafter, we will use a new class called `ViewTreeObserver`
for quick peeks and hiding:

```
@Override
protected void onCreate(Bundle savedInstanceState) {
    super.onCreate(savedInstanceState);
    setContentView(R.layout.activity_main);
    setAmbientEnabled();

  //Initialise the arraylist
  drawer_itemArrayList = initializeScreenSystem();
  mSelectedScreen = 0;

  // Initialize here for content to first screen.
  ...

  // Main Wearable Drawer Layout that holds all the content
  mWearableDrawerLayout = (WearableDrawerLayout)
  findViewById(R.id.drawer_layout);

  // Top Navigation Drawer
  mWearableNavigationDrawer =
          (WearableNavigationDrawer)
  findViewById(R.id.top_navigation_drawer);

  mWearableNavigationDrawer.setAdapter(new NavigationAdapter(this));

  // Bottom Action Drawer
```

```
    mWearableActionDrawer =
            (WearableActionDrawer)
    findViewById(R.id.bottom_action_drawer);

    mWearableActionDrawer.setOnMenuItemClickListener(this);

    // Temporarily peeks the navigation and action drawers to ensure
    the user is aware of them.
    ViewTreeObserver observer =
    mWearableDrawerLayout.getViewTreeObserver();
    observer.addOnGlobalLayoutListener(new
    ViewTreeObserver.OnGlobalLayoutListener() {
        @Override
        public void onGlobalLayout() {
            mWearableDrawerLayout.getViewTreeObserver()
            .removeOnGlobalLayoutListener(this);
            mWearableDrawerLayout.peekDrawer(Gravity.TOP);
            mWearableDrawerLayout.peekDrawer(Gravity.BOTTOM);
        }
    });
}
```

We have seen the array initialization code after the setAmbientEnabled() method. We need to write the associated method to it. Inside the method, initialize a list of DrawerItem, loop through the items, and get the title and icon, as follows:

```
private ArrayList<DrawerItem> initializeScreenSystem() {
    ArrayList<DrawerItem> screens = new ArrayList<DrawerItem>();
    String[] FragmentArrayNames =
    getResources().getStringArray(R.array.screens);

    for (int i = 0; i < FragmentArrayNames.length; i++) {
        String planet = FragmentArrayNames[i];
        int FragmentResourceId =
                getResources().getIdentifier(planet, "array",
                getPackageName());
        String[] fragmentInformation =
        getResources().getStringArray(FragmentResourceId);

        screens.add(new DrawerItem(
                fragmentInformation[0],   // Name
                fragmentInformation[1]));
    }

    return screens;
}
```

In the `FragmentArrayNames` instance, we are searching for a string array, which needs to be created. In the `res/values` folder, create the `arrays.xml` file and the following set of arrays:

```xml
<?xml version="1.0" encoding="utf-8"?>
<resources>

    <string-array name="screens">
        <item>water</item>
        <item>heart</item>
        <item>step</item>
    </string-array>

    <string-array name="water">
       // drawer item title
        <item>Drink water</item>
      // drawer item icon
        <item>water_bottle_flat</item>
    </string-array>

    <string-array name="heart">
        <item>Heart Beat</item>
        <item>ic_heart_icon</item>
    </string-array>

    <string-array name="step">
        <item>Step Counter</item>
        <item>ic_step_icon</item>
    </string-array>

</resources>
```

For understanding, let's take the water item in the `screens` string array. We add a title and exact icon name to the items for the water fragment screen, and it is the same for the other array items.

For the `WearableActionDrawer` item, we need to have the actions configured before we continue writing the Java logic. We need to add the `menu` XML file into the `res/menu` directory. Let's call the file `drawer_menu.xml`:

```xml
<?xml version="1.0" encoding="utf-8"?>

<menu xmlns:android="http://schemas.android.com/apk/res/android">
    <item android:id="@+id/menu_about"
            android:icon="@drawable/ic_info"
            android:title="About"/>
    <item android:id="@+id/menu_helathtips"
```

```
        android:icon="@drawable/ic_info"
        android:title="Health tips" />
    <item android:id="@+id/menu_calarie"
        android:icon="@drawable/ic_info"
        android:title="Calories chart" />
</menu>
```

We have to override onMenuItemclicklistener and fireup an action when the user clicks on any of the menu items:

```
@Override
public boolean onMenuItemClick(MenuItem menuItem) {
    Log.d(TAG, "onMenuItemClick(): " + menuItem);

    final int itemId = menuItem.getItemId();

    String toastMessage = "";

    switch (itemId) {
        case R.id.menu_about:
            toastMessage =
            drawer_itemArrayList.get(mSelectedScreen).getName();
            break;
        case R.id.menu_helathtips:
            toastMessage =
            drawer_itemArrayList.get(mSelectedScreen).getName();
            break;
        case R.id.menu_volume:
            toastMessage =
            drawer_itemArrayList.get(mSelectedScreen).getName();
            break;
    }


    mWearableDrawerLayout.closeDrawer(mWearableActionDrawer);

    if (toastMessage.length() > 0) {
        Toast toast = Toast.makeText(
                getApplicationContext(),
                toastMessage,
                Toast.LENGTH_SHORT);
        toast.show();
        return true;
    } else {
        return false;
    }
}
```

At the point when `actionDrawer` popsup from the bottom of the Wear device, we will close the navigation `drawerlayout` to have the activity drawer perceivably. When a user clicks on the item, we will show the fragment screen name, which we made in array XML.

Now for the last piece of `MainActivity`. Let's compose the `NavigationDrawer` adapter to attach the frame when it is switched.

Creating a navigation drawer adapter

Make a class that extends to `WearableNavigationDrawerAdapter` and override the following methods:

```
private final class NavigationAdapter
        extends WearableNavigationDrawer
        .WearableNavigationDrawerAdapter {

@Override
public String getItemText(int i) {
    return null;
}

@Override
public Drawable getItemDrawable(int i) {
    return null;
}

@Override
public void onItemSelected(int i) {

}

@Override
public int getCount() {
    return 0;
}

}
```

After this, make a constructor that gets the context as a parameter, and, inside the `getCount` method, return the `drawer_item ArrayList` with its size.

In the `onItemSelected` method, based on the position parameter, we can switch the fragments:

```
@Override
public void onItemSelected(int position) {
```

```
        Log.d(TAG, "WearableNavigationDrawerAdapter.onItemSelected(): " +
        position);
        mSelectedScreen = position;

        if(position==0) {
            final DrinkWaterFragment drinkWaterFragment = new
            DrinkWaterFragment();
            getFragmentManager()
                    .beginTransaction()
                    .replace(R.id.content_frame, drinkWaterFragment)
                    .commit();
        }

    }
```

Regardless, we have to write the fragments. Hang on for some time. In `getItemText`, we can restore the name from `draweritem ArrayList` as follows:

```
@Override
public String getItemText(int pos) {
    return drawer_itemArrayList.get(pos).getName();
}
```

To fetch the drawable icon and set the icon, we are going to use the following custom method:

```
@Override
public Drawable getItemDrawable(int pos) {
    String navigationIcon =
    drawer_itemArrayList.get(pos).getNavigationIcon();

    int drawableNavigationIconId =
            getResources().getIdentifier(navigationIcon, "drawable",
            getPackageName());

    return mContext.getDrawable(drawableNavigationIconId);
}
```

Yes, we have completed the `MainActivity` code. Let's put all the methods together and see the complete `MainActivity` code:

```
public class MainActivity extends WearableActivity  implements
        WearableActionDrawer.OnMenuItemClickListener{


    private static final String TAG = "MainActivity";
    private WearableDrawerLayout mWearableDrawerLayout;
    private WearableNavigationDrawer mWearableNavigationDrawer;
```

```java
private WearableActionDrawer mWearableActionDrawer;
private ArrayList<DrawerItem> drawer_itemArrayList;
private int mSelectedScreen;

private DrinkWaterFragment mDrinkFragment;


@Override
protected void onCreate(Bundle savedInstanceState) {
    super.onCreate(savedInstanceState);
    setContentView(R.layout.activity_main);
    setAmbientEnabled();
    drawer_itemArrayList = initializeScreenSystem();
    mSelectedScreen = 0;

    // Initialize content to first screen.
    mDrinkFragment = new DrinkWaterFragment();
    FragmentManager fragmentManager = getFragmentManager();
    fragmentManager.beginTransaction().replace(R.id.content_frame,
    mDrinkFragment).commit();

    // Main Wearable Drawer Layout that holds all the content
    mWearableDrawerLayout = (WearableDrawerLayout)
    findViewById(R.id.drawer_layout);

    // Top Navigation Drawer
    mWearableNavigationDrawer =
            (WearableNavigationDrawer)
    findViewById(R.id.top_navigation_drawer);

    mWearableNavigationDrawer.setAdapter(new
    NavigationAdapter(this));

    // Bottom Action Drawer
    mWearableActionDrawer =
            (WearableActionDrawer)
            findViewById(R.id.bottom_action_drawer);

    mWearableActionDrawer.setOnMenuItemClickListener(this);

    // Temporarily peeks the navigation and action drawers to
    ensure the user is aware of them.
    ViewTreeObserver observer =
    mWearableDrawerLayout.getViewTreeObserver();
    observer.addOnGlobalLayoutListener(new
    ViewTreeObserver.OnGlobalLayoutListener() {
        @Override
        public void onGlobalLayout() {
```

```
                    mWearableDrawerLayout.getViewTreeObserver()
                    .removeOnGlobalLayoutListener(this);
                    mWearableDrawerLayout.peekDrawer(Gravity.TOP);
                    mWearableDrawerLayout.peekDrawer(Gravity.BOTTOM);
            }
        });
    }

    private ArrayList<DrawerItem> initializeScreenSystem() {
        ArrayList<DrawerItem> screens = new ArrayList<DrawerItem>();
        String[] FragmentArrayNames =
        getResources().getStringArray(R.array.screens);

        for (int i = 0; i < FragmentArrayNames.length; i++) {
            String planet = FragmentArrayNames[i];
            int FragmentResourceId =
                    getResources().getIdentifier(planet, "array",
                    getPackageName());
            String[] fragmentInformation =
            getResources().getStringArray(FragmentResourceId);

            screens.add(new DrawerItem(
                    fragmentInformation[0],    // Name
                    fragmentInformation[1]));
        }

        return screens;
    }

    @Override
    public boolean onMenuItemClick(MenuItem menuItem) {
        Log.d(TAG, "onMenuItemClick(): " + menuItem);

        final int itemId = menuItem.getItemId();

        String toastMessage = "";

        switch (itemId) {
            case R.id.menu_about:
                toastMessage =
                drawer_itemArrayList.get(mSelectedScreen).getName();
                break;
            case R.id.menu_helathtips:
                toastMessage =
                drawer_itemArrayList.get(mSelectedScreen).getName();
                break;
            case R.id.menu_volume:
                toastMessage =
```

```
            drawer_itemArrayList.get(mSelectedScreen).getName();
            break;
    }


    mWearableDrawerLayout.closeDrawer(mWearableActionDrawer);

    if (toastMessage.length() > 0) {
        Toast toast = Toast.makeText(
                getApplicationContext(),
                toastMessage,
                Toast.LENGTH_SHORT);
        toast.show();
        return true;
    } else {
        return false;
    }
}


private final class NavigationAdapter
        extends
WearableNavigationDrawer.WearableNavigationDrawerAdapter {

    private final Context mContext;

    public NavigationAdapter(Context context) {
        mContext = context;
    }

    @Override
    public int getCount() {
        return drawer_itemArrayList.size();
    }

    @Override
    public void onItemSelected(int position) {
        Log.d(TAG,
        "WearableNavigationDrawerAdapter.onItemSelected():"
        + position);
        mSelectedScreen = position;

        if(position==0) {
            final DrinkWaterFragment drinkWaterFragment = new
            DrinkWaterFragment();
            getFragmentManager()
                    .beginTransaction()
                    .replace(R.id.content_frame,
```

```
                    drinkWaterFragment)
                    .commit();

        }
        }

    }

    @Override
    public String getItemText(int pos) {
        return drawer_itemArrayList.get(pos).getName();
    }

    @Override
    public Drawable getItemDrawable(int pos) {
        String navigationIcon =
        drawer_itemArrayList.get(pos).getNavigationIcon();

        int drawableNavigationIconId =
                getResources().getIdentifier(navigationIcon,
                "drawable", getPackageName());

        return mContext.getDrawable(drawableNavigationIconId);
    }
}
    @Override
public void onEnterAmbient(Bundle ambientDetails) {
    super.onEnterAmbient(ambientDetails);
}

@Override
public void onUpdateAmbient() {
    super.onUpdateAmbient();
}

@Override
public void onExitAmbient() {
    super.onExitAmbient();
}
}
```

Creating fragments

We need to create a fragment called `DrinkWaterFragment`. In this fragment, we will handle starting and ending the `WaterDrink` reminder.

Create a new layout file in the `res/layout` directory, create the `drink_water_fragment.xml` file, and add two buttons inside `boxinsetlayout`, as follows:

```xml
<?xml version="1.0" encoding="utf-8"?>
<android.support.wearable.view.BoxInsetLayout
xmlns:android="http://schemas.android.com/apk/res/android"
    xmlns:app="http://schemas.android.com/apk/res-auto"
    xmlns:tools="http://schemas.android.com/tools"
    android:id="@+id/container"
    android:layout_width="match_parent"
    android:layout_height="match_parent"
    tools:context="com.packt.upbeat.fragments.HeartRateFragment"
    tools:deviceIds="wear">

    <LinearLayout
        android:layout_width="match_parent"
        android:layout_height="wrap_content"
        android:layout_gravity="center"
        android:orientation="horizontal">

        <android.support.v7.widget.AppCompatButton
            android:id="@+id/start"
            android:layout_width="match_parent"
            android:layout_height="50dp"
            android:layout_margin="10dp"
            android:layout_weight="1"
            android:background="@drawable/button_background"
            android:elevation="5dp"
            android:gravity="center"
            android:text="Start"
            android:textAllCaps="true"
            android:textColor="@color/white"
            android:textStyle="bold" />

        <android.support.v7.widget.AppCompatButton
            android:id="@+id/stop"
            android:layout_width="match_parent"
            android:layout_height="50dp"
            android:layout_margin="10dp"
            android:layout_weight="1"
            android:background="@drawable/button_background"
            android:elevation="5dp"
            android:gravity="center"
            android:text="Stop"
            android:textAllCaps="true"
            android:textColor="@color/white"
            android:textStyle="bold" />
```

```
        </LinearLayout>
    </android.support.wearable.view.BoxInsetLayout>
```

You can use the normal `Button` class as well, but, in this project, we are going to use `AppCompatButton` for material design illustrations such as elevation and other features. We have to customize the button background selector. Create a file inside the `drawable` directory and call it `button_background.xml`. Add the following code:

```
<?xml version="1.0" encoding="utf-8"?>
<selector xmlns:android="http://schemas.android.com/apk/res/android">
    <item android:drawable="@color/colorPrimaryDark"
    android:state_pressed="true"/>
    <item android:drawable="@color/grey" android:state_focused="true"/>
    <item android:drawable="@color/colorPrimary"/>
</selector>
```

Inside `DrinkWaterFragment`, instantiate the buttons, and attach a listener to both the buttons for starting and stopping the service of, as follows:

```
public class DrinkWaterFragment extends Fragment {


    private AppCompatButton mStart;
    private AppCompatButton mStop;


    public DrinkWaterFragment() {
        // Required empty public constructor
    }


    @Override
    public View onCreateView(LayoutInflater inflater, ViewGroup
    container,
                             Bundle savedInstanceState) {
        // Inflate the layout for this fragment
        View rootView = inflater.inflate(R.layout.drink_water_fragment,
        container, false);

        mStart = (AppCompatButton) rootView.findViewById(R.id.start);
        mStop = (AppCompatButton) rootView.findViewById(R.id.stop);
        mStart.setOnClickListener(new View.OnClickListener() {
            @Override
            public void onClick(View v) {

    }
});
```

```
mStop.setOnClickListener(new View.OnClickListener() {
    @Override
    public void onClick(View v) {

    }
});

    return rootView;
    }
}
```

It is now time to write a `BroadcastReceiver` class inside the services package with the notification configured inside. Let's call the class `WaterReminderReceiver` and extend it to `BroadcastReciever`. Override the `onReceive` method and, whenever `AlarmManager` triggers this receiver will receive the data, we will see a notification:

```
public class WaterReminderReceiver extends BroadcastReceiver {
    public static final String CONTENT_KEY = "contentText";

    @Override
    public void onReceive(Context context, Intent intent) {
        Intent intent2 = new Intent(context, MainActivity.class);
        intent.addFlags(Intent.FLAG_ACTIVITY_CLEAR_TOP);
        PendingIntent pendingIntent =
        PendingIntent.getActivity(context, 0, intent2,
                PendingIntent.FLAG_ONE_SHOT);

        Uri defaultSoundUri =
        RingtoneManager.getDefaultUri(RingtoneManager.TYPE_ALARM);

        NotificationCompat.Builder notificationBuilder =
        (NotificationCompat.Builder) new
        NotificationCompat.Builder(context)
                .setAutoCancel(true)
                //Automatically delete the notification
                .setSmallIcon(R.drawable.water_bottle_flat)
                //Notification icon
                .setContentIntent(pendingIntent)
                .setContentTitle("Time to hydrate")
                .setContentText("Drink a glass of water now")
                .setCategory(Notification.CATEGORY_REMINDER)
                .setPriority(Notification.PRIORITY_HIGH)
                .setSound(defaultSoundUri);

        NotificationManagerCompat notificationManager =
        NotificationManagerCompat.from(context);
        notificationManager.notify(0, notificationBuilder.build());
```

```
        Toast.makeText(context, "Repeating Alarm Received",
        Toast.LENGTH_SHORT).show();
    }
}
```

Register this receiver in the manifest inside the application tag scope:

```
<receiver android:name=".services.WaterReminderReceiver"
android:process=":remote" />
```

Now, in the `DrinkWaterFragment` in the Start button's `onClickListener` start the `AlarmManager` service and register the broadcast receiver as well:

```
mStart.setOnClickListener(new View.OnClickListener() {
    @Override
    public void onClick(View v) {
        Intent intent = new Intent(getActivity(),
        WaterReminderReceiver.class);
        PendingIntent sender =
        PendingIntent.getBroadcast(getActivity(),
                0, intent, 0);

        // We want the alarm to go off 5 seconds from now.
        long firstTime = SystemClock.elapsedRealtime();
        firstTime += 5 * 1000;

        // Schedule the alarm!
        AlarmManager am = (AlarmManager)
        getActivity().getSystemService(ALARM_SERVICE);
        am.setRepeating(AlarmManager.ELAPSED_REALTIME_WAKEUP,
                firstTime, 5 * 1000, sender);

        //DOZE MODE SUPPORT        am.setAndAllowWhileIdle(AlarmManager
        .ELAPSED_REALTIME_WAKEUP, firstTime, sender);

        // Tell the user about what we did.
        if (mToast != null) {
            mToast.cancel();
        }
        mToast = Toast.makeText(getActivity(),
        "Subscribed to water alarm",
                Toast.LENGTH_LONG);
        mToast.show();
    }
});
```

Now, for debugging purposes, we will be giving a delay of 15 seconds, though you can change the delay based on your needs and application use case. We also make sure that when a Wear device goes to the doze mode the water reminder still firesup the doze-supported method to start the alarm manager.

To stop the water reminder service when we click the **Stop** button, onClickListener will stop the alarm manager that is running in background and shows a quick toast for the unsubscription of water reminder:

```
mStop.setOnClickListener(new View.OnClickListener() {
        @Override
        public void onClick(View v) {
            // Create the same intent, and thus a matching
            IntentSender, for the one that was scheduled.
            Intent intent = new Intent(getActivity(),
            WaterReminderReceiver.class);
            PendingIntent sender =
            PendingIntent.getBroadcast(getActivity(),
                    0, intent, 0);

            // And cancel the alarm.
            AlarmManager am = (AlarmManager)
            getActivity().getSystemService(ALARM_SERVICE);
            am.cancel(sender);

            // Tell the user about what we did.
            if (mToast != null) {
                mToast.cancel();
            }
            mToast = Toast.makeText(getActivity(), "Unsubscribed
            from water reminder",
                    Toast.LENGTH_LONG);
            mToast.show();
        }
    });
```

The complete fragment class code is as follows:

```
public class DrinkWaterFragment extends Fragment {

    private AppCompatButton mStart;
    private AppCompatButton mStop;
    private Toast mToast;

    public DrinkWaterFragment() {
        // Required empty public constructor
    }
```

```
@Override
public View onCreateView(LayoutInflater inflater, ViewGroup
container,
                        Bundle savedInstanceState) {
    // Inflate the layout for this fragment
    View rootView = inflater.inflate(R.layout.drink_water_fragment,
    container, false);

    mStart = (AppCompatButton) rootView.findViewById(R.id.start);
    mStop = (AppCompatButton) rootView.findViewById(R.id.stop);

    mStart.setOnClickListener(new View.OnClickListener() {
        @Override
        public void onClick(View v) {
            Intent intent = new Intent(getActivity(),
            WaterReminderReceiver.class);
            PendingIntent sender =
            PendingIntent.getBroadcast(getActivity(),
                    0, intent, 0);

            // We want the alarm to go off 5 seconds from now.
            long firstTime = SystemClock.elapsedRealtime();
            firstTime += 5 * 1000;

            // Schedule the alarm!
            AlarmManager am = (AlarmManager)
            getActivity().getSystemService(ALARM_SERVICE);
            am.setRepeating(AlarmManager.ELAPSED_REALTIME_WAKEUP,
                    firstTime, 5 * 1000, sender);
            //DOZE MODE SUPPORT
            am.setAndAllowWhileIdle
            (AlarmManager.ELAPSED_REALTIME_WAKEUP,
            firstTime, sender);

            // Tell the user about what we did.
            if (mToast != null) {
                mToast.cancel();
            }
            mToast = Toast.makeText(getActivity(), "Subscribed to
            water alarm",
                    Toast.LENGTH_LONG);
            mToast.show();
        }
    });

    mStop.setOnClickListener(new View.OnClickListener() {
        @Override
        public void onClick(View v) {
```

```
            // Create the same intent,
            and thus a matching IntentSender, for
            // the one that was scheduled.
            Intent intent = new Intent(getActivity(),
            WaterReminderReceiver.class);
            PendingIntent sender =
            PendingIntent.getBroadcast(getActivity(),
                    0, intent, 0);

            // And cancel the alarm.
            AlarmManager am = (AlarmManager)
            getActivity().getSystemService(ALARM_SERVICE);
            am.cancel(sender);

            // Tell the user about what we did.
            if (mToast != null) {
                mToast.cancel();
            }
            mToast = Toast.makeText(getActivity(), "Unsubscribed
            from water reminder",
                    Toast.LENGTH_LONG);
            mToast.show();
        }
    });

    return rootView;
    }
}
```

This fragment is attached to the `MainActivity NavigationDrawerAdapter`. Ensure you attach this fragment as the default fragment by attaching the fragment in the `onCreate` method as the `MainActivity`:

```
@Override
protected void onCreate(Bundle savedInstanceState) {

...
// Initialize content to first screen.
mDrinkFragment = new DrinkWaterFragment();
FragmentManager fragmentManager = getFragmentManager();
fragmentManager.beginTransaction().replace(R.id.content_frame,
mDrinkFragment).commit();

}
```

We have successfully completed the drink water reminder feature for the app. Now, let's build the heartrate detection through the optical heart rate sensors.

Create a fragment inside the fragments package and call that `HeartRateFragment.java`. In the `res/layout` directory refactor, create XML, or create a new layout file and call that `heart_rate_fragment.xml`, and add the following code:

```xml
<?xml version="1.0" encoding="utf-8"?>
<android.support.wearable.view.BoxInsetLayout
xmlns:android="http://schemas.android.com/apk/res/android"
    xmlns:tools="http://schemas.android.com/tools"
    android:id="@+id/container"
    android:layout_width="match_parent"
    android:layout_height="match_parent"
    tools:context="com.packt.upbeat.fragments.HeartRateFragment"
    tools:deviceIds="wear">

    <LinearLayout xmlns:app="http://schemas.android.com/apk/res-auto"
        android:layout_width="match_parent"
        android:layout_height="match_parent"
        android:layout_gravity="center"
        android:orientation="horizontal">

        <ImageView
        android:layout_weight="1"
        android:src="@drawable/ic_heart_icon"
        android:layout_width="match_parent"
        android:layout_height="match_parent" />

        <TextView
            android:id="@+id/heart_rate"
            android:layout_width="match_parent"
            android:layout_height="match_parent"
            android:layout_gravity="center"
            android:layout_weight="1"
            android:gravity="left|center"
            android:hint="Reading"
            android:textColor="@color/colorAccent"
            android:textColorHint="#eaeaea"
            android:textSize="20sp"
            android:textStyle="bold" />

    </LinearLayout>
</android.support.wearable.view.BoxInsetLayout>
```

Instead of a static `ImageView`, as shown in the previous code, how about adding a dynamic heartbeat for the pulse rate returned by a sensor? We will create a custom scale animation based on the open source project at `https://github.com/scottyab/HeartBeatView`. Let's create a class inside the utils package and call that `HeartBeatView.java`, which extends to `AppCompatImageView`. Before we get started on working on the `CustomView`, we have to set its styleables, as follows, which helps to manage the custom values we pass. Create a file in `res/values` and call it `heartbeat_attrs.xml`. `Styleable` defines the attributes and properties for the custom view; for instance, when the custom view needs a scale factor, we can define it as shown in the following code example:

```xml
<?xml version="1.0" encoding="utf-8"?>
<resources>

    <declare-styleable name="HeartBeatView">
        <attr name="scaleFactor" format="float" />
        <attr name="duration" format="integer" />
    </declare-styleable>

</resources>
```

Create the `HeartBeatView` class by extending the `AppCompatImageView` class. We need to create constructors that belong to the `AppCompatImageView` class, as follows:

```java
public class HeartBeatView extends AppCompatImageView{

    public HeartBeatView(Context context) {
        super(context);
    }

    public HeartBeatView(Context context, AttributeSet attrs) {
        super(context, attrs);
    }

    public HeartBeatView(Context context, AttributeSet attrs, int
    defStyleAttr) {
        super(context, attrs, defStyleAttr);
    }
}
```

Let's create instances for configuring the scale factor and animation duration. Add the following instances to the global scope of the class:

```java
private static final float DEFAULT_SCALE_FACTOR = 0.2f;
private static final int DEFAULT_DURATION = 50;
private Drawable heartDrawable;
```

```
private boolean heartBeating = false;

float scaleFactor = DEFAULT_SCALE_FACTOR;
float reductionScaleFactor = -scaleFactor;
int duration = DEFAULT_DURATION;
```

We need a vector graphic in a heart shape, and that needs to be created in the `drawable` folder. Create a file, call it `heart_red_24dp.xml`, and add the following code:

```
<vector xmlns:android="http://schemas.android.com/apk/res/android"
    android:width="24dp"
    android:height="24dp"
    android:viewportHeight="24.0"
    android:viewportWidth="24.0">
    <path
        android:fillColor="#FFFF0000"
        android:pathData="M12,21.35l-1.45,-1.32C5.4,15.36 2,12.28 2,8.5
        2,5.42 4.42,3 7.5,3c1.74,0 3.41,0.81 4.5,2.09C13.09,3.81
        14.76,3 16.5,3 19.58,3 22,5.42 22,8.5c0,3.78 -3.4,6.86
        -8.55,11.54L12,21.35z"/>
</vector>
```

To initialize the vector graphic, we will utilize the `drawable` instance and access the vector graphic:

```
private void init() {
    //make this not mandatory
    heartDrawable = ContextCompat.getDrawable(getContext(),
    R.drawable.ic_heart_red_24dp);
    setImageDrawable(heartDrawable);

}
```

Using the `styleable` that we created in the `res/values` directory, we can populate the scale factor and animation duration. Using the `TypedArray` class instance, we can obtain the attributes, as follows:

```
(Context context, AttributeSet attrs) {
    TypedArray a = context.getTheme().obtainStyledAttributes(
            attrs,
            R.styleable.HeartBeatView,
            0, 0
    );
    try {
        scaleFactor = a.getFloat(R.styleable.HeartBeatView_scaleFactor,
        DEFAULT_SCALE_FACTOR);
        reductionScaleFactor = -scaleFactor;
        duration = a.getInteger(R.styleable.HeartBeatView_duration,
```

```
            DEFAULT_DURATION);


        } finally {
            a.recycle();
        }


    }
```

We will require three methods, namely `toggle()`, `start()`, and `stop()` to initialize the animation:

```java
/**
 * toggles current heat beat state
 */
public void toggle() {
    if (heartBeating) {
        stop();
    } else {
        start();
    }
}

/**
 * Starts the heat beat/pump animation
 */
public void start() {
    heartBeating = true;
    animate().scaleXBy(scaleFactor).scaleYBy(scaleFactor)
    .setDuration(duration).setListener(scaleUpListener);
}

/**
 * Stops the heat beat/pump animation
 */
public void stop() {
    heartBeating = false;
    clearAnimation();
}
```

Now, to set the duration based on BPM, we will simply assign bpm to the duration using the `Math.round` operation for a 3-digit floating point:

```
public void setDurationBasedOnBPM(int bpm) {
    if (bpm > 0) {
        duration = Math.round((milliInMinute / bpm) / 3f);
    }
}
```

To check whether the `heartBeat` animation has started and for the duration of the animation, we have to write the following two methods:

```
public boolean isHeartBeating() {
    return heartBeating;
}

public int getDuration() {
    return duration;
}
```

Using this pointer, we will assign the duration and the scale factor:

```
public void setDuration(int duration) {
    this.duration = duration;
}

public float getScaleFactor() {
    return scaleFactor;
}

public void setScaleFactor(float scaleFactor) {
    this.scaleFactor = scaleFactor;
    reductionScaleFactor = -scaleFactor;
}
```

Finally, we will write two animation listener methods for `ScaleUpAnimation` and `ScaleDownAnimation`. We will write a method of the type `Animator.AnimatorListener` and we will increase the scale in `ScaleUpListener` and decrease it in `ScaleDownListener` as follows:

```
//Scale up animation
private final Animator.AnimatorListener scaleUpListener = new
Animator.AnimatorListener() {

    @Override
    public void onAnimationStart(Animator animation) {
    }
```

```
    @Override
    public void onAnimationRepeat(Animator animation) {

    }

    @Override
    public void onAnimationEnd(Animator animation) {
        //we ignore heartBeating as we want to ensure the heart is
        reduced back to original size
        animate().scaleXBy(reductionScaleFactor)
        .scaleYBy(reductionScaleFactor).setDuration(duration)
        .setListener(scaleDownListener);
    }

    @Override
    public void onAnimationCancel(Animator animation) {

    }
};


//Scale down animation
private final Animator.AnimatorListener scaleDownListener = new
Animator.AnimatorListener() {

    @Override
    public void onAnimationStart(Animator animation) {
    }

    @Override
    public void onAnimationRepeat(Animator animation) {
    }

    @Override
    public void onAnimationEnd(Animator animation) {
        if (heartBeating) {
            //duration twice as long for the upscale
            animate().scaleXBy(scaleFactor).scaleYBy(scaleFactor)
            .setDuration(duration * 2).setListener(scaleUpListener);
        }
    }

    @Override
    public void onAnimationCancel(Animator animation) {
    }
};
```

The completed custom view class is as follows:

```java
public class HeartBeatView extends AppCompatImageView {

    private static final String TAG = "HeartBeatView";

    private static final float DEFAULT_SCALE_FACTOR = 0.2f;
    private static final int DEFAULT_DURATION = 50;
    private Drawable heartDrawable;

    private boolean heartBeating = false;

    float scaleFactor = DEFAULT_SCALE_FACTOR;
    float reductionScaleFactor = -scaleFactor;
    int duration = DEFAULT_DURATION;

    public HeartBeatView(Context context) {
        super(context);
        init();
    }

    public HeartBeatView(Context context, AttributeSet attrs) {
        super(context, attrs);
        populateFromAttributes(context, attrs);
        init();
    }

    public HeartBeatView(Context context, AttributeSet attrs, int
    defStyleAttr) {
        super(context, attrs, defStyleAttr);
        populateFromAttributes(context, attrs);
        init();
    }

    private void init() {
        //make this not mandatory
        heartDrawable = ContextCompat.getDrawable(getContext(),
        R.drawable.ic_heart_red_24dp);
        setImageDrawable(heartDrawable);

    }

    private void populateFromAttributes(Context context, AttributeSet
    attrs) {
        TypedArray a = context.getTheme().obtainStyledAttributes(
                attrs,
                R.styleable.HeartBeatView,
                0, 0
```

```
    );
    try {
        scaleFactor =
        a.getFloat(R.styleable.HeartBeatView_scaleFactor,
        DEFAULT_SCALE_FACTOR);
        reductionScaleFactor = -scaleFactor;
        duration = a.getInteger(R.styleable.HeartBeatView_duration,
        DEFAULT_DURATION);


    } finally {
        a.recycle();
    }

}


/**
 * toggles current heat beat state
 */
public void toggle() {
    if (heartBeating) {
        stop();
    } else {
        start();
    }
}

/**
 * Starts the heat beat/pump animation
 */
public void start() {
    heartBeating = true;
    animate().scaleXBy(scaleFactor).scaleYBy(scaleFactor)
    .setDuration(duration).setListener(scaleUpListener);
}

/**
 * Stops the heat beat/pump animation
 */
public void stop() {
    heartBeating = false;
    clearAnimation();
}

/**
 * is the heart currently beating
 *
```

```
 * @return
 */
public boolean isHeartBeating() {
    return heartBeating;
}

public int getDuration() {
    return duration;
}


private static final int milliInMinute = 60000;

/**
 * set the duration of the beat based on the beats per minute
 *
 * @param bpm (positive int above 0)
 */
public void setDurationBasedOnBPM(int bpm) {
    if (bpm > 0) {
        duration = Math.round((milliInMinute / bpm) / 3f);
    }
}


public void setDuration(int duration) {
    this.duration = duration;
}

public float getScaleFactor() {
    return scaleFactor;
}

public void setScaleFactor(float scaleFactor) {
    this.scaleFactor = scaleFactor;
    reductionScaleFactor = -scaleFactor;
}

private final Animator.AnimatorListener scaleUpListener = new
Animator.AnimatorListener() {

    @Override
    public void onAnimationStart(Animator animation) {
    }

    @Override
    public void onAnimationRepeat(Animator animation) {
```

```
    }

    @Override
    public void onAnimationEnd(Animator animation) {
        //we ignore heartBeating as we want to ensure the heart is
        reduced back to original size
        animate().scaleXBy(reductionScaleFactor)
        .scaleYBy(reductionScaleFactor).setDuration(duration)
        .setListener(scaleDownListener);
    }

    @Override
    public void onAnimationCancel(Animator animation) {


    }
};


private final Animator.AnimatorListener scaleDownListener = new
Animator.AnimatorListener() {

    @Override
    public void onAnimationStart(Animator animation) {
    }

    @Override
    public void onAnimationRepeat(Animator animation) {
    }

    @Override
    public void onAnimationEnd(Animator animation) {
        if (heartBeating) {
            //duration twice as long for the upscale
            animate().scaleXBy(scaleFactor).scaleYBy(scaleFactor)
            .setDuration(duration * 2)
            .setListener(scaleUpListener);
        }
    }

    @Override
    public void onAnimationCancel(Animator animation) {
    }
};

}
```

In `heart_rate_fragment.xml`, replace the `imageview` code with the new custom view created. Give it a unique ID:

```
<com.packt.upbeat.utils.HeartBeatView
    android:id="@+id/heartbeat"
    android:layout_width="match_parent"
    android:layout_height="match_parent"
    android:layout_marginLeft="25dp"
    android:layout_weight="1" />
```

Create the `HeartRateFragment` class, extend it to the `Fragment` class, and implement it to `SensorEventListener`. `SensorEventListener` will monitor all the sensor updates and return the changed result:

```
public class HeartRateFragment extends Fragment implements
SensorEventListener
{
    public HeartRateFragment() {
    // Required empty public constructor
    }


    @Override
    public View onCreateView(LayoutInflater inflater, ViewGroup container,
    Bundle savedInstanceState) {
        // Inflate the layout for this fragment
        View rootView = inflater.inflate(R.layout.heart_rate_fragment,
        container, false);

    return rootView;

    }
}
```

Within in the global scope of the program, instantiate the following necessary components:

```
private BoxInsetLayout mContainerView;
private TextView mTextView;
private HeartBeatView heartbeat;
private Sensor mHeartRateSensor;
private SensorManager mSensorManager;
private Integer currentValue = 0;
private static final String TAG = "HeartRateFragment";
private static final int SENSOR_PERMISSION_CODE = 123;
```

In the `onCreateView` method, map all the visual components added into the XML file. It is good practice to initialize the sensors in `onCreateView` for facing less number of NPEs:

```
heartbeat = (HeartBeatView)rootView.findViewById(R.id.heartbeat);

mContainerView = (BoxInsetLayout)rootView.findViewById(R.id.container);
mTextView = (TextView)rootView.findViewById(R.id.heart_rate);
mSensorManager =
((SensorManager)getActivity().getSystemService(SENSOR_SERVICE));
mHeartRateSensor = mSensorManager.getDefaultSensor(Sensor.TYPE_HEART_RATE);
```

Accessing the `HeartRate` sensor needs permission in the manifest. In Wear 2.0, we need to set the runtime permission as well. So, register the body sensor permission in the manifest file:

```
<uses-permission android:name="android.permission.WAKE_LOCK" />
<uses-permission android:name="android.permission.BODY_SENSORS" />
<uses-permission android:name="android.permission.RECEIVE_BOOT_COMPLETED"/>
```

For runtime permission, we need to request the access for `BODY_SENSORS` and then override the `onRequestPermissionresul()` method. The following code illustrates the runtime permission model:

```
//Requesting permission
  private void requestSensorPermission() {
      if (ContextCompat.checkSelfPermission(getActivity(),
      Manifest.permission.BODY_SENSORS) ==
      PackageManager.PERMISSION_GRANTED)
          return;

      if (ActivityCompat.shouldShowRequestPermissionRationale
      (getActivity(), Manifest.permission.BODY_SENSORS)) {
          //If the user has denied the permission previously your
          code will come to this block
          //Here you can explain why you need this permission
          //Explain here why you need this permission
      }
      //And finally ask for the permission
      ActivityCompat.requestPermissions(getActivity(), new String[]
      {Manifest.permission.BODY_SENSORS}, SENSOR_PERMISSION_CODE);
  }

  //This method will be called when the user will tap
  on allow or deny
  @Override
  public void onRequestPermissionsResult(int requestCode,
  @NonNull String[] permissions, @NonNull int[] grantResults) {
```

```
            //Checking the request code of our request
            if (requestCode == SENSOR_PERMISSION_CODE) {

                //If permission is granted
                if (grantResults.length > 0 && grantResults[0] ==
                PackageManager.PERMISSION_GRANTED) {
                    //Displaying a toast
                    Toast.makeText(getActivity(), "Permission granted now
                    you can read the storage", Toast.LENGTH_LONG).show();
                } else {
                    //Displaying another toast if permission is not granted
                    Toast.makeText(getActivity(), "Oops you just denied the
                    permission", Toast.LENGTH_LONG).show();
                }
            }
        }
    }
```

Now, call `requestSensorPermission()` in the `onCreateView()` method:

```
@Override
public View onCreateView(LayoutInflater inflater, ViewGroup container,
                        Bundle savedInstanceState) {
    // Inflate the layout for this fragment
    View rootView = inflater.inflate(R.layout.heart_rate_fragment,
    container, false);


  //other components.

    requestSensorPermission();

    return rootView;
}
```

In the `onSensorChanged()` method, using the `sensorevent` object from the argument, we will now get the `HeartRate`. The following code fetches the type of sensor and its return values. Later, inside the for loop, we can set the `HeartBeat` animation duration and its toggle method to start the animation:

```
@Override
public void onSensorChanged(SensorEvent sensorEvent) {

    if(sensorEvent.sensor.getType() == Sensor.TYPE_HEART_RATE &&
    sensorEvent.values.length > 0) {

        for(Float value : sensorEvent.values) {
```

```
        int newValue = Math.round(value);

        if(currentValue != newValue) {
            currentValue = newValue;

            mTextView.setText(currentValue.toString());
            heartbeat.setDurationBasedOnBPM(currentValue);
            heartbeat.toggle();
        }

    }

  }
}
```

In the `onStart` and `onDestroy` methods, register the `HeartRate` sensor and unregister the sensor:

```
@Override
public void onStart() {
    super.onStart();
    if (mHeartRateSensor != null) {
        Log.d(TAG, "HEART RATE SENSOR NAME: " +
        mHeartRateSensor.getName() + " TYPE: " +
        mHeartRateSensor.getType());
        mSensorManager.unregisterListener(this, this.mHeartRateSensor);
        boolean isRegistered = mSensorManager.registerListener(this,
        this.mHeartRateSensor, SensorManager.SENSOR_DELAY_FASTEST);
        Log.d(TAG, "HEART RATE LISTENER REGISTERED: " + isRegistered);
    } else {
        Log.d(TAG, "HEART RATE SENSOR NOT READY");
    }
}


@Override
public void onDestroy() {
    super.onDestroy();
    mSensorManager.unregisterListener(this);
    Log.d(TAG, "SENSOR UNREGISTERED");
}
```

The complete `HeartRateFragment` class code is as follows:

```java
public class HeartRateFragment extends Fragment implements
SensorEventListener {

    private BoxInsetLayout mContainerView;
    private TextView mTextView;
    private HeartBeatView heartbeat;
    private Sensor mHeartRateSensor;
    private SensorManager mSensorManager;
    private Integer currentValue = 0;
    private static final String TAG = "HeartRateFragment";
    private static final int SENSOR_PERMISSION_CODE = 123;

    private GoogleApiClient mGoogleApiClient;

    public HeartRateFragment() {
        // Required empty public constructor
    }


    @Override
    public View onCreateView(LayoutInflater inflater, ViewGroup
    container,
                             Bundle savedInstanceState) {
        // Inflate the layout for this fragment
        View rootView = inflater.inflate(R.layout.heart_rate_fragment,
        container, false);


        heartbeat = (HeartBeatView) rootView.findViewById
        (R.id.heartbeat);

        mContainerView = (BoxInsetLayout) rootView.findViewById
        (R.id.container);
        mTextView = (TextView) rootView.findViewById(R.id.heart_rate);
        mSensorManager = ((SensorManager) getActivity()
        .getSystemService(SENSOR_SERVICE));
        mHeartRateSensor = mSensorManager.getDefaultSensor
        (Sensor.TYPE_HEART_RATE);


        mGoogleApiClient = new GoogleApiClient.Builder(getActivity())
        .addApi(Wearable.API).build();
        mGoogleApiClient.connect();


        requestSensorPermission();
```

```
        return rootView;
    }

@Override
public void onStart() {
    super.onStart();
    if (mHeartRateSensor != null) {
        Log.d(TAG, "HEART RATE SENSOR NAME: " +
        mHeartRateSensor.getName() + " TYPE: " +
        mHeartRateSensor.getType());
        mSensorManager.unregisterListener(this,
        this.mHeartRateSensor);
        boolean isRegistered = mSensorManager.registerListener
        (this, this.mHeartRateSensor,
        SensorManager.SENSOR_DELAY_FASTEST);
        Log.d(TAG, "HEART RATE LISTENER REGISTERED: " +
        isRegistered);
    } else {
        Log.d(TAG, "HEART RATE SENSOR NOT READY");
    }
}


@Override
public void onDestroy() {
    super.onDestroy();
    mSensorManager.unregisterListener(this);
    Log.d(TAG, "SENSOR UNREGISTERED");
}


@Override
public void onSensorChanged(SensorEvent sensorEvent) {

    if(sensorEvent.sensor.getType() == Sensor.TYPE_HEART_RATE &&
    sensorEvent.values.length > 0) {

        for(Float value : sensorEvent.values) {

            int newValue = Math.round(value);

            if(currentValue != newValue) {
                currentValue = newValue;

                mTextView.setText(currentValue.toString());
                heartbeat.setDurationBasedOnBPM(currentValue);
                heartbeat.toggle();
            }
```

```
            }

        }
    }

    @Override
    public void onAccuracyChanged(Sensor sensor, int i) {
        Log.d(TAG, "ACCURACY CHANGED: " + i);
    }


    //Requesting permission
    private void requestSensorPermission() {
        if (ContextCompat.checkSelfPermission(getActivity(),
        Manifest.permission.BODY_SENSORS) ==
        PackageManager.PERMISSION_GRANTED)
            return;

        if (ActivityCompat.shouldShowRequestPermissionRationale
        (getActivity(), Manifest.permission.BODY_SENSORS)) {
            //If the user has denied the permission previously your
            code will come to this block
            //Here you can explain why you need this permission
            //Explain here why you need this permission
        }
        //And finally ask for the permission
        ActivityCompat.requestPermissions(getActivity(), new String[]
        {Manifest.permission.BODY_SENSORS}, SENSOR_PERMISSION_CODE);
    }

    //This method will be called when the user will tap
    on allow or deny
    @Override
    public void onRequestPermissionsResult(int requestCode, @NonNull
    String[] permissions, @NonNull int[] grantResults) {

        //Checking the request code of our request
        if (requestCode == SENSOR_PERMISSION_CODE) {

            //If permission is granted
            if (grantResults.length > 0 && grantResults[0] ==
            PackageManager.PERMISSION_GRANTED) {
                //Displaying a toast
                Toast.makeText(getActivity(), "Permission granted now
                 you can read the storage", Toast.LENGTH_LONG).show();
            } else {
                //Displaying another toast if permission is not granted
                Toast.makeText(getActivity(), "Oops you just denied the
```

```
                permission", Toast.LENGTH_LONG).show();
            }
        ]
    }
}
```

Finally, in the `MainActivity` navigation adapter, we can attach `HeartRateFragment` for the second index value. In the `onItemSelected` method, add the following code change:

```
if(position==0) {
    final DrinkWaterFragment drinkWaterFragment = new
    DrinkWaterFragment();
    getFragmentManager()
            .beginTransaction()
            .replace(R.id.content_frame, drinkWaterFragment)
            .commit();

}else if(position == 1){
    final HeartRateFragment sectionFragment = new HeartRateFragment();
    getFragmentManager()
            .beginTransaction()
            .replace(R.id.content_frame, sectionFragment)
            .commit();
}
```

We have to build one more screen for the Step counter. Create a new fragment in the fragments package. Create another layout xml file and call it `step_counter_fragment.xml`. The user interface for this chapter's scope is just two text fields,.

as follows:

```
<?xml version="1.0" encoding="utf-8"?>
<android.support.wearable.view.BoxInsetLayout
xmlns:android="http://schemas.android.com/apk/res/android"
    xmlns:tools="http://schemas.android.com/tools"
    android:id="@+id/container"
    android:layout_width="match_parent"
    android:layout_height="match_parent"
    tools:context="com.packt.upbeat.fragments.HeartRateFragment"
    tools:deviceIds="wear">

    <LinearLayout
        xmlns:android="http://schemas.android.com/apk/res/android"
        android:layout_width="match_parent"
        android:layout_height="match_parent"
        android:orientation="vertical">
```

```xml
<TextView
    android:text="@string/steps"
    android:layout_width="wrap_content"
    android:layout_height="wrap_content"
    android:layout_gravity="center_horizontal|bottom"
    android:textSize="24sp"
    android:layout_weight="1"/>

<TextView
    android:id="@+id/steps"
    android:layout_gravity="center_horizontal|top"
    android:layout_width="wrap_content"
    android:layout_height="wrap_content"
    android:textSize="24sp"
    android:layout_weight="1"/>

    </LinearLayout>
</android.support.wearable.view.BoxInsetLayout>
```

In the fragment class created, we have one simple text view to display the `StepCounter` data. To keep the step counter sensor running in the background, we need to create a service class and attach the sensor data to the service class. Before constructing the Fragments, let's take care of the services that the step counter will require.

We will write an observer to receive a specific datatype and, in the UI, we will use a Handler thread to receive the data. Let's create a class called `EventReceiver`, which listens to the sensor changes. We are going to use the `BlockingQueue` class for java, which waits for the queue to get empty. Using a separate thread, with the help of `ThreadPoolExecutor`, we can detect the events. Find the complete class, as follows:

```java
public class EventReceiver {
    private static ConcurrentMap<Class<?>, ConcurrentMap<Reporter,
    String>> events
        = new ConcurrentHashMap<Class<?>, ConcurrentMap<Reporter,
    String>>();

    private static BlockingQueue<Runnable> queue = new
    LinkedBlockingQueue<Runnable>();
    private static ExecutorService executorService = new
    ThreadPoolExecutor(1, 10, 30, TimeUnit.SECONDS, queue);

    public static void register(Class<?> event, Reporter reporter) {
        if (null == event || null == reporter)
            return;

        events.putIfAbsent(event, new ConcurrentHashMap<Reporter,
        String>());
```

```
            events.get(event).putIfAbsent(reporter, "");
    }

    public static void remove(Class<?> event, Reporter reporter) {
        if (null == event || null == reporter)
            return;

        if (!events.containsKey(event))
            return;

        events.get(event).remove(reporter);
    }

    public static void notify(final Object event) {
        if (null == event)
            return;

        if (!events.containsKey(event.getClass()))
            return;

        for (final Reporter m : events.get(event.getClass()).keySet()) {
            executorService.execute(new Runnable() {
                @Override
                public void run() {
                    m.notifyEvent(event);
                }
            });
        }
    }

}
```

We will register the notification from the `Reporter` interface and remove it whenever it is not necessary. The reporter interface is as follows:

```
public interface Reporter {
    public void notifyEvent(Object o);
}
```

We need a `BroadcastReceiver` class to receive the notification from the service:

```
public class AlarmNotification extends BroadcastReceiver {

    private static final String TAG = "AlarmNotification";

    @Override
    public void onReceive(Context context, Intent intent) {
        Log.d(TAG, "alarm fired");
```

```
        context.startService(new Intent(context,
        WearStepService.class));
    }
}
```

Let's get started on writing the service class. Create a class that extends to the Service class and implement it to `SensorEventListener` with the following instances:

```
public class WearStepService extends Service implements SensorEventListener
{

    public static final String TAG = "WearStepService";
    private static final long THREE_MINUTES = 3 * 60 * 1000;
    private static final String STEP_COUNT_PATH = "/step-count";
    private static final String STEP_COUNT_KEY = "step-count";
    private SensorManager sensorManager;
    private Sensor countSensor;


}
```

Create a method for getting the sensor manager:

```
private void getSensorManager() {
    if (null != sensorManager)
        return;

    Log.d(TAG, "getSensorManager");
    sensorManager = (SensorManager)
    getSystemService(Context.SENSOR_SERVICE);
    registerCountSensor();
}
```

The following method gets the Step counter sensor from the Wear device:

```
private void getCountSensor() {
    if (null != countSensor)
        return;

    Log.d(TAG, "getCountSensor");
    countSensor =
    sensorManager.getDefaultSensor(Sensor.TYPE_STEP_COUNTER);
    registerCountSensor();
}
```

To register the step count sensor, we will use the sensor manager class and, using the `registerListener` method, we will register the sensor:

```
private void registerCountSensor() {
```

```
    if (countSensor == null)
        return;

    Log.d(TAG, "sensorManager.registerListener");
    sensorManager.registerListener(this, countSensor,
    SensorManager.SENSOR_DELAY_UI);
}
```

To set up the `BroadcastReceiver` alarm, we created an `AlarmNotification` `BroadcastReceiver` earlier. Using the following method, we can register the class:

```
private void setAlarm() {
    Log.d(TAG, "setAlarm");

    Intent intent = new Intent(this, AlarmNotification.class);
    PendingIntent pendingIntent =
    PendingIntent.getBroadcast(this.getApplicationContext(),
    234324243, intent, 0);
    AlarmManager alarmManager = (AlarmManager)
    getSystemService(ALARM_SERVICE);
    long firstRun = System.currentTimeMillis() + THREE_MINUTES;
    alarmManager.setInexactRepeating(AlarmManager.RTC_WAKEUP, firstRun,
    THREE_MINUTES, pendingIntent);
}
```

Whenever a sensor gives a variation in the sensor event data, using the `onSensorChanged` method, we can fireup the notification in the background:

```
@Override
public void onSensorChanged(SensorEvent event) {
    if (event.sensor.getType() == Sensor.TYPE_STEP_COUNTER)
        StepsTaken.updateSteps(event.values.length);
    Log.d(TAG, "onSensorChanged: steps count is" +
    event.values.length);
    updateNotification();
}
```

For the `UpdateNotification` method, we will use the `NotificationCompat.Builder` class to construct the notification:

```
private void updateNotification() {
    // Create a notification builder that's compatible with platforms
    >= version 4
    NotificationCompat.Builder builder =
            new NotificationCompat.Builder(getApplicationContext());

    // Set the title, text, and icon
    builder.setContentTitle(getString(R.string.app_name))
```

```
            .setSmallIcon(R.drawable.ic_step_icon);

    builder.setContentText("steps: " + StepsTaken.getSteps());

    // Get an instance of the Notification Manager
    NotificationManager notifyManager = (NotificationManager)
            getSystemService(Context.NOTIFICATION_SERVICE);

    // Build the notification and post it
    notifyManager.notify(0, builder.build());
}
```

Override the `onStartCommand` method, and initialize the sensor manager and step counter sensor, as follows:

```
@Override
public int onStartCommand(Intent intent, int flags, int startId) {
    Log.d(TAG, "onStartCommand");

    getSensorManager();
    getCountSensor();

    return super.onStartCommand(intent, flags, startId);
}
```

Whenever there is an accuracy change in the sensor readings, we can fireup the notification as follows:

```
@Override
public void onAccuracyChanged(Sensor sensor, int accuracy) {
    // drop these messages
    updateNotification();

}
```

We need to register the service and broadcast receiver in the manifest within the application tag:

```
<service android:name=".services.WearStepService" />

<receiver android:name=".services.AlarmNotification">
    <intent-filter>
        <action android:name="android.intent.action.BOOT_COMPLETED"/>
    </intent-filter>
</receiver>
```

We have to create a simple `StepsTaken` Logic for the service to start fresh every day. We will create a serializable class, and, using the Calendar instance if the day is over, we will initialise the steps from zero:

```
public class StepsTaken implements Serializable {

    private static int steps = 0;
    private static long lastUpdateTime = 0L;
    private static final String TAG = "StepsTaken";

    public static void updateSteps(int stepsTaken) {
        steps += stepsTaken;

        // today
        Calendar tomorrow = new GregorianCalendar();
        tomorrow.setTimeInMillis(lastUpdateTime);
        // reset hour, minutes, seconds and millis
        tomorrow.set(Calendar.HOUR_OF_DAY, 0);
        tomorrow.set(Calendar.MINUTE, 0);
        tomorrow.set(Calendar.SECOND, 0);
        tomorrow.set(Calendar.MILLISECOND, 0);

        // next day
        tomorrow.add(Calendar.DAY_OF_MONTH, 1);

        Calendar now = Calendar.getInstance();

        if (now.after(tomorrow)) {
            Log.d(TAG, "I think it's tomorrow, resetting");
            steps = stepsTaken;
        }

        lastUpdateTime = System.currentTimeMillis();
    }

    public static int getSteps() {
        return steps;
    }
}
```

The complete code for `WearStepService` class is given below.

```
public class WearStepService extends Service implements SensorEventListener
{

    public static final String TAG = "WearStepService";
    private static final long THREE_MINUTES = 3 * 60 * 1000;
    private SensorManager sensorManager;
```

```
    private Sensor countSensor;


    GoogleApiClient mGoogleApiClient;

    @Override
    public void onCreate() {
        super.onCreate();
        Log.d(TAG, "onCreate");
        setAlarm();
    }

    @Override
    public int onStartCommand(Intent intent, int flags, int startId) {
        Log.d(TAG, "onStartCommand");

        getSensorManager();
        getCountSensor();
        getGoogleClient();

        return super.onStartCommand(intent, flags, startId);
    }

    @Override
    public IBinder onBind(Intent intent) {
        return null;
    }

    private void getGoogleClient() {
        if (null != mGoogleApiClient)
            return;

        Log.d(TAG, "getGoogleClient");
        mGoogleApiClient = new GoogleApiClient.Builder(this)
                .addApi(Wearable.API)
                .build();
        mGoogleApiClient.connect();
    }

    /**
     * if the countSensor is null, try initializing it, and try
     registering it with sensorManager
     */
    private void getCountSensor() {
        if (null != countSensor)
            return;

        Log.d(TAG, "getCountSensor");
```

```
        countSensor = sensorManager
        .getDefaultSensor(Sensor.TYPE_STEP_COUNTER);
        registerCountSensor();
    }

    /**
     * if the countSensor exists, then try registering
     */
    private void registerCountSensor() {
        if (countSensor == null)
            return;

        Log.d(TAG, "sensorManager.registerListener");
        sensorManager.registerListener(this, countSensor,
        SensorManager.SENSOR_DELAY_UI);
    }

    /**
     * if the sensorManager is null, initialize it, and try registering
     the countSensor
     */
    private void getSensorManager() {
        if (null != sensorManager)
            return;

        Log.d(TAG, "getSensorManager");
        sensorManager = (SensorManager)
        getSystemService(Context.SENSOR_SERVICE);
        registerCountSensor();
    }

    private void setAlarm() {
        Log.d(TAG, "setAlarm");

        Intent intent = new Intent(this, AlarmNotification.class);
        PendingIntent pendingIntent = PendingIntent.getBroadcast
        (this.getApplicationContext(), 234324243, intent, 0);
        AlarmManager alarmManager = (AlarmManager)
        getSystemService(ALARM_SERVICE);
        long firstRun = System.currentTimeMillis() + THREE_MINUTES;
        alarmManager.setInexactRepeating(AlarmManager.RTC_WAKEUP,
        firstRun, THREE_MINUTES, pendingIntent);
    }

    @Override
    public void onSensorChanged(SensorEvent event) {
        if (event.sensor.getType() == Sensor.TYPE_STEP_COUNTER)
            StepsTaken.updateSteps(event.values.length);
```

```
          Log.d(TAG, "onSensorChanged: steps count is" +
          event.values.length);
//          sendToPhone();
          sendData();
          updateNotification();
      }

      private void sendData(){

          if (mGoogleApiClient == null)
              return;

          // use the api client to send the heartbeat value to our
          handheld
          final PendingResult<NodeApi.GetConnectedNodesResult> nodes =
          Wearable.NodeApi.getConnectedNodes(mGoogleApiClient);
          nodes.setResultCallback(new
          ResultCallback<NodeApi.GetConnectedNodesResult>() {
              @Override
              public void onResult(NodeApi.GetConnectedNodesResult
              result) {
                  final List<Node> nodes = result.getNodes();
                  final String path = "/stepcount";
                  String Message = StepsTaken.getSteps()+"";

                  for (Node node : nodes) {
                      Log.d(TAG, "SEND MESSAGE TO HANDHELD: " + Message);
                      node.getDisplayName();
                      byte[] data =
                      Message.getBytes(StandardCharsets.UTF_8);
                      Wearable.MessageApi.sendMessage(mGoogleApiClient,
                      node.getId(), path, data);
                  }
              }
          });
      }

      private void updateNotification() {
          // Create a notification builder that's compatible with
          platforms >= version 4
          NotificationCompat.Builder builder =
                  new NotificationCompat.Builder
                  (getApplicationContext());

          // Set the title, text, and icon
          builder.setContentTitle(getString(R.string.app_name))
                  .setSmallIcon(R.drawable.ic_step_icon);
```

```
        builder.setContentText("steps: " + StepsTaken.getSteps());

        // Get an instance of the Notification Manager
        NotificationManager notifyManager = (NotificationManager)
                getSystemService(Context.NOTIFICATION_SERVICE);

        // Build the notification and post it
        notifyManager.notify(0, builder.build());
    }

    @Override
    public void onAccuracyChanged(Sensor sensor, int accuracy) {
        // drop these messages
        updateNotification();


    }
}
```

Now, finally, we shall update the step counter code in the fragment. In the
StepCounterFragment, we need to implement the Reporter interface and create the
handler instance and the textview instance, as follows:

```
public class StepCounterFragment extends Fragment implements Reporter {


    private TextView tv;
    private Handler handler = new Handler();


    public StepCounterFragment() {
        // Required empty public constructor
    }

    @Override
    public View onCreateView(LayoutInflater inflater, ViewGroup container,
                        Bundle savedInstanceState) {
        // Inflate the layout for this fragment
        View rootView = inflater.inflate(R.layout.step_counter_fragment,
        container, false); return rootView;
    }
}
```

Now, in the oncreateView method, register WearStepService and connect the step
counter textview to the xml tag we created, as follows:

```
getActivity().startService(new Intent(getActivity(),
WearStepService.class));
```

```
tv = (TextView)rootView.findViewById(R.id.steps);
tv.setText(String.valueOf(StepsTaken.getSteps()));
```

When we implement the Reporter interface, we need to override a `NotifyEvent` method, as follows:

```
@Override
public void notifyEvent(final Object o) {
    handler.post(new Runnable() {
        @Override
        public void run() {
            if (o instanceof StepsTaken)
                tv.setText(String.valueOf(StepsTaken.getSteps()));
        }
    });

}
```

In the `onResume` and `onPause` methods of the Fragment lifecycle, register and remove the event receiver observer for the `StepsTaken` class we wrote earlier:

```
@Override
public void onResume() {
    EventReceiver.register(StepsTaken.class, this);
    super.onResume();
}

@Override
public void onPause() {
    EventReceiver.remove(StepsTaken.class, this);
    super.onPause();
}
```

The complete fragment class looks as follows:

```
public class StepCounterFragment extends Fragment implements Reporter {


    private TextView tv;
    private Handler handler = new Handler();


    public StepCounterFragment() {
        // Required empty public constructor
    }


    @Override
```

```java
public View onCreateView(LayoutInflater inflater, ViewGroup
container,
                            Bundle savedInstanceState) {
    // Inflate the layout for this fragment
    View rootView =
    inflater.inflate(R.layout.step_counter_fragment,
    container, false);

    getActivity().startService(new Intent(getActivity(),
    WearStepService.class));


    tv = (TextView)rootView.findViewById(R.id.steps);
    tv.setText(String.valueOf(StepsTaken.getSteps()));

    return rootView;
}

@Override
public void notifyEvent(final Object o) {
    handler.post(new Runnable() {
        @Override
        public void run() {
            if (o instanceof StepsTaken)
                tv.setText(String.valueOf(StepsTaken.getSteps()));
        }
    });

}

@Override
public void onResume() {
    EventReceiver.register(StepsTaken.class, this);
    super.onResume();
}

@Override
public void onPause() {
    EventReceiver.remove(StepsTaken.class, this);
    super.onPause();
}

}
```

We have successfully completed the Stepcounter feature. Do not forget to attach this fragment in the MainActivity navigation adapter with the next index value:

```java
if(position==0) {
```

```
        final DrinkWaterFragment drinkWaterFragment = new DrinkWaterFragment();
        getFragmentManager()
                .beginTransaction()
                .replace(R.id.content_frame, drinkWaterFragment)
                .commit();

    }else if(position == 1){
        final HeartRateFragment sectionFragment = new HeartRateFragment();
        getFragmentManager()
                .beginTransaction()
                .replace(R.id.content_frame, sectionFragment)
                .commit();
    }else if(position == 2){
        final StepCounterFragment stepCounterFragment = new
        StepCounterFragment();
        getFragmentManager()
                .beginTransaction()
                .replace(R.id.content_frame, stepCounterFragment)
                .commit();
    }
```

The app will look as follows once you complete all the fragments.

The following screenshot illustrates the drink water fragment that helps the user start the reminder service:

The following screenshot illustrates how the water reminder can be started and stop through these buttons:

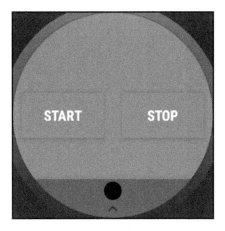

This image illustrates how to read the `HeartRate` sensor as well as the `HeartBeat` Animation we created:

This image illustrates the step counter screen, which shows the step counts:

Summary

In this chapter, we have studied the sensors, battery utilizations, and the best practices. We have additionally looked at how the doze mode helps to spare the battery and spare the CPU cycles. You have learned how to construct a material design Wear application using ideas such as making fragments for Wear devices, working with `WearableNavigationLayout`, working with `WearableActionDrawer`, working with services and `BroadcastReceivers`, and working with sensors, such as the optical `HeartRate` sensor and step counter, the notifications for the step counter and drink water services, and runtime permissions for the body sensors.

In the next chapter, we anticipate making this application more solid by using more elements and functionalities.

5
Measuring Your Wellness and Syncing Collected Sensor Data

In the previous chapter, we built a Wear app that reminds us to drink water, and has the ability to check step counts and heart pulse rate through embedded Wear sensors. The ideology of Wear and mobile applications is to have better glanceability and never miss anything important, and project Upbeat seems, with few functionalities, to have a strong place in a user's wrist and pocket. The current functionality of the upbeat Wear app is limited to displaying the data received from the sensors. In this chapter, we are going to sweeten the app with interoperability with Wear and mobile apps. We will persist all transmitted data through `RealmDB`. We will fire a notification from a mobile to Wear to start the app for checking the pulse rate. We will have a list of health tips and food calorie cards in the Wear application.

In this chapter, we will explore the following:

- Collecting Wear sensor's data
- Processing received data to find calories and distance
- `WearableListenerService` and Messaging API
- Sending data to a Wear app from the mobile app
- `RealmDB` integration
- `WearableRecyclerview` with `CardView`

Collecting Wear sensors' data

Collecting sensor data from Wear devices needs a communication mechanism, and the Wearable DataLayer API, which is part of the Google Play services, plays a major role in the communication process. We will deeply explore the communication process in later lessons but in this chapter, we need to receive the sensor data in the mobile app. We have created the project, which already has a mobile module and the simple plain old Hello World boilerplate code. We will work on the mobile module once we set our data sending mechanism from the Wear app. Let's start with the step count sensor in the services package of the Wear module and go to the `WearStepService` class. We have built this service to fire a notification and listen to the step counter data. Now, with the help of `GoogleApiClient` and the Wear Messaging API, we need to send the data to the mobile app.

In the `WearStepService` class, instantiate `GoogleApiClient` in the global scope of the class:

```
GoogleApiClient mGoogleApiClient;
```

In `onStartCommand`, call a method that initializes `mGoogleApiClient`:

```
@Override
public int onStartCommand(Intent intent, int flags, int startId) {
    Log.d(TAG, "onStartCommand");

    getSensorManager();
    getCountSensor();
    getGoogleClient();

    return super.onStartCommand(intent, flags, startId);
}
```

For initializing `GoogleClient`, we will utilize the builder pattern of the `GoogleClient` and we need to add `Wearable.API`. Later, we can connect `GoogleClient` using the `connect()` method followed by the `build()` method of the builder:

```
private void getGoogleClient() {
    if (null != mGoogleApiClient)
        return;

    Log.d(TAG, "getGoogleClient");
    mGoogleApiClient = new GoogleApiClient.Builder(this)
            .addApi(Wearable.API)
            .build();
    mGoogleApiClient.connect();
}
```

Within the `WearStepService` class, we shall override the `onBind` method that belongs to the `IBinder` interface. We can use it for client interactions through the remote service as follows:

```
@Override
public IBinder onBind(Intent intent) {
    return null;
}
```

We are returning null, since after sending data, we don't want anything to return. If we want certain information to be returned, then we can return the `IBinder` instance as follows:

```
private final IBinder mBinder = new LocalBinder();

@Override
public IBinder onBind(Intent intent) {
  return mBinder;
}
```

To be able to send data to mobile devices, we need two DataLayer API mechanisms, which are Wearable Node and Message API. We will get the connected nodes using Node API. Using the Messaging API, we will fire the data to a particular path, and on the receiver end, we should listen to that path for fetching the data.

In the Node API, we will have the `Resultcallback` class, which returns a list of `ConnectedNodes`, and we have to implement the `onResult` method that has the ability to return the list of connected nodes. We can fire the message to all the connected nodes or for one that is connected. We can get the connected node's name using `getDisplayname` of the node class, as follows:

```
node.getDisplayName();
```

For now, we will use the Node and Message API and send the data to the connected node:

```
private void sendData(){

    if (mGoogleApiClient == null)
        return;

    // use the api client to send the heartbeat value to our handheld
    final PendingResult<NodeApi.GetConnectedNodesResult> nodes =
    Wearable.NodeApi.getConnectedNodes(mGoogleApiClient);
    nodes.setResultCallback(new
    ResultCallback<NodeApi.GetConnectedNodesResult>() {
        @Override
```

```
public void onResult(NodeApi.GetConnectedNodesResult result) {
    final List<Node> nodes = result.getNodes();
    final String path = "/stepcount";
    String Message = StepsTaken.getSteps()+"";

    for (Node node : nodes) {
        Log.d(TAG, "SEND MESSAGE TO HANDHELD: " + Message);
        node.getDisplayName();
        byte[] data = Message.getBytes(StandardCharsets.UTF_8);
        Wearable.MessageApi.sendMessage(mGoogleApiClient,
        node.getId(), path, data);
    }
}
});
}
```

In the previous method, we will use the `PendingResults` class to retreive the connected node's results. After we receive the list of connected nodes, we can fire Messages using the `wearableMessageApi` class. Do not forget to send and receive the data to the same path.

Completed WearStepService class

The complete `WearStepService` class code is as follows:

```
public class WearStepService extends Service implements SensorEventListener
{

    public static final String TAG = "WearStepService";
    private static final long THREE_MINUTES = 3 * 60 * 1000;
    private static final String STEP_COUNT_PATH = "/step-count";
    private static final String STEP_COUNT_KEY = "step-count";
    private SensorManager sensorManager;
    private Sensor countSensor;


    GoogleApiClient mGoogleApiClient;

    @Override
    public void onCreate() {
        super.onCreate();
        Log.d(TAG, "onCreate");
        setAlarm();
    }

    @Override
    public int onStartCommand(Intent intent, int flags, int startId) {
```

```
        Log.d(TAG, "onStartCommand");

        getSensorManager();
        getCountSensor();
        getGoogleClient();

        return super.onStartCommand(intent, flags, startId);
    }

    @Override
    public IBinder onBind(Intent intent) {
        return null;
    }

    private void getGoogleClient() {
        if (null != mGoogleApiClient)
            return;

        Log.d(TAG, "getGoogleClient");
        mGoogleApiClient = new GoogleApiClient.Builder(this)
                .addApi(Wearable.API)
                .build();
        mGoogleApiClient.connect();
    }

    /**
     * if the countSensor is null, try initializing it, and try
     registering it with sensorManager
     */
    private void getCountSensor() {
        if (null != countSensor)
            return;

        Log.d(TAG, "getCountSensor");
        countSensor =
        sensorManager.getDefaultSensor(Sensor.TYPE_STEP_COUNTER);
        registerCountSensor();
    }

    /**
     * if the countSensor exists, then try registering
     */
    private void registerCountSensor() {
        if (countSensor == null)
            return;

        Log.d(TAG, "sensorManager.registerListener");
        sensorManager.registerListener(this, countSensor,
```

```
        SensorManager.SENSOR_DELAY_UI);
    }

    /**
     * if the sensorManager is null, initialize it, and try registering
      the countSensor
     */
    private void getSensorManager() {
        if (null != sensorManager)
            return;

        Log.d(TAG, "getSensorManager");
        sensorManager = (SensorManager)
        getSystemService(Context.SENSOR_SERVICE);
        registerCountSensor();
    }

    private void setAlarm() {
        Log.d(TAG, "setAlarm");

        Intent intent = new Intent(this, AlarmNotification.class);
        PendingIntent pendingIntent =
        PendingIntent.getBroadcast(this.getApplicationContext(),
        234324243, intent, 0);
        AlarmManager alarmManager = (AlarmManager)
        getSystemService(ALARM_SERVICE);
        long firstRun = System.currentTimeMillis() + THREE_MINUTES;
        alarmManager.setInexactRepeating(AlarmManager.RTC_WAKEUP,
        firstRun, THREE_MINUTES, pendingIntent);
    }

    @Override
    public void onSensorChanged(SensorEvent event) {
        if (event.sensor.getType() == Sensor.TYPE_STEP_COUNTER)
            StepsTaken.updateSteps(event.values.length);
        Log.d(TAG, "onSensorChanged: steps count is" +
        event.values.length);
//        sendToPhone();
        sendData();
        updateNotification();
    }

    private void sendData(){

        if (mGoogleApiClient == null)
            return;

        // use the api client to send the heartbeat value to our
```

```
    handheld
    final PendingResult<NodeApi.GetConnectedNodesResult> nodes =
    Wearable.NodeApi.getConnectedNodes(mGoogleApiClient);
    nodes.setResultCallback(new
    ResultCallback<NodeApi.GetConnectedNodesResult>() {
        @Override
        public void onResult(NodeApi.GetConnectedNodesResult
        result) {
            final List<Node> nodes = result.getNodes();
            final String path = "/stepcount";
            String Message = StepsTaken.getSteps()+"";

            for (Node node : nodes) {
                Log.d(TAG, "SEND MESSAGE TO HANDHELD: " + Message);
                node.getDisplayName();
                byte[] data =
                Message.getBytes(StandardCharsets.UTF_8);
                Wearable.MessageApi.sendMessage(mGoogleApiClient,
                node.getId(), path, data);
            }
        }
    });
}

private void updateNotification() {
    // Create a notification builder that's compatible with
    // platforms >= version 4
    NotificationCompat.Builder builder =
            new NotificationCompat.Builder
            (getApplicationContext());

    // Set the title, text, and icon
    builder.setContentTitle(getString(R.string.app_name))
            .setSmallIcon(R.drawable.ic_step_icon);

    builder.setContentText("steps: " + StepsTaken.getSteps());

    // Get an instance of the Notification Manager
    NotificationManager notifyManager = (NotificationManager)
            getSystemService(Context.NOTIFICATION_SERVICE);

    // Build the notification and post it
    notifyManager.notify(0, builder.build());
}

@Override
public void onAccuracyChanged(Sensor sensor, int accuracy) {
    // drop these messages
```

```
            updateNotification();

        }
    }
```

We are successfully firing a message to the mentioned path. Now, let's see how to retrieve the message from the Wear device. Inside the mobile module, create the additional package for code readability. We are going to name the packages models, services, and utils.

It's time to create a class that extends to WearableListenerService with the override method onMessageReceived. Create a class called StepListener and extend it to WearableListenerService; the code is as follows:

```
    public class StepListner extends WearableListenerService {

        private static final String TAG = "StepListner";

        @Override
        public void onMessageReceived(MessageEvent messageEvent) {
            if (messageEvent.getPath().equals("/stepcount")) {
                final String message = new String(messageEvent.getData());
                Log.v(TAG, "Message path received from wear is: " +
                messageEvent.getPath());
                Log.v(TAG, "Message received on watch is: " + message);

                // Broadcast message to wearable activity for display
                Intent messageIntent = new Intent();
                messageIntent.setAction(Intent.ACTION_SEND);
                messageIntent.putExtra("message", message);
                LocalBroadcastManager.getInstance(this)
                .sendBroadcast(messageIntent);
            }
            else {
                super.onMessageReceived(messageEvent);
            }
        }
    }
```

Registering the earlier service class in the manifest with the same path that the data sent over, the following code illustrates Wearable DATA_CHANGED and MESSAGE_RECEIVED actions with the data path:

```
    <service android:name=".services.StepListner">
        <intent-filter>
            <action android:name=
            "com.google.android.gms.wearable.DATA_CHANGED" />
            <action android:name=
```

```
        "com.google.android.gms.wearable.MESSAGE_RECEIVED" />

        <data
            android:host="*"
            android:pathPrefix="/stepcount"
            android:scheme="wear" />
    </intent-filter>
</service>
```

The `Steplistner` class is completed; we can use this class to process the data further. In the `steplistener` class, we are registering the `localbroadcast` receiver class to send the received data across the scope of the broadcast receiver. Before we build a UI, we shall receive all the data in the mobile app in `MainActivity`. Write an inner class to read through the steps received:

```
public class StepReceiver extends BroadcastReceiver {
        @Override
        public void onReceive(Context context, Intent intent) {
            String message = intent.getStringExtra("message");
            Log.v("steps", "Main activity received message: " +
            message);

        // Shows the step counts received by the wearlistener //service
            mSteps.setText("Steps:"+ message);
            int value = Integer.valueOf(message);
        }
    }
```

Register the class in the `oncreate` method with the following code:

```
// Register the local broadcast receiver
IntentFilter StepFilter = new IntentFilter(Intent.ACTION_SEND);
StepReceiver StepReceiver = new StepReceiver();
LocalBroadcastManager.getInstance(this).registerReceiver(StepReceiver,
StepFilter);
```

We are successfully collecting the step counter data. Let's do the same process to collect the pulse rate. We will persist the step counts and, later, we will have a live stream of the heart pulse rate transmitting over the connected nodes in real time.

Switching back to the Wear module

Switch the project scope to the Wear module and select `HeartRateFragment` to instantiate a `GoogleClient` object:

```
private GoogleApiClient mGoogleApiClient;
```

Initialize the `GoogleClient` instance within the `oncreate` method as follows:

```
mGoogleApiClient = new GoogleApiClient.Builder(getActivity()).
addApi(Wearable.API).
build();

mGoogleApiClient.connect();
```

Write a method that sends the pulse rate count to the connected nodes, as we did earlier for the step counter:

```
private void sendMessageToHandheld(final String message) {

    if (mGoogleApiClient == null)
        return;

    // use the api client to send the heartbeat value to our handheld
    final PendingResult<NodeApi.GetConnectedNodesResult> nodes =
    Wearable.NodeApi.getConnectedNodes(mGoogleApiClient);
    nodes.setResultCallback(new
    ResultCallback<NodeApi.GetConnectedNodesResult>() {
        @Override
        public void onResult(NodeApi.GetConnectedNodesResult result) {
            final List<Node> nodes = result.getNodes();
            final String path = "/heartRate";

            for (Node node : nodes) {
                Log.d(TAG, "SEND MESSAGE TO HANDHELD: " + message);

                byte[] data = message.getBytes(StandardCharsets.UTF_8);
                Wearable.MessageApi.sendMessage(mGoogleApiClient,
                node.getId(), path, data);
            }
        }
    });
}
```

Call the method inside the `onSensorchanged` callback with the BPM count received from the sensor event trigger:

```
sendMessageToHandheld(currentValue.toString());
```

Switch to the mobile project scope. We need one more `WearableListenerService` class to talk with the heart rate data:

```
public class HeartListener extends WearableListenerService {

    @Override
    public void onMessageReceived(MessageEvent messageEvent) {


    }

}
```

Register a `localbroadcast` event inside the `onMessageReceived` callback to receive the data in activities. The complete listener class code is as follows:

```
public class HeartListener extends WearableListenerService {

    @Override
    public void onMessageReceived(MessageEvent messageEvent) {
        if (messageEvent.getPath().equals("/heartRate")) {
            final String message = new String(messageEvent.getData());
            Log.v("pactchat", "Message path received on watch is: " +
            messageEvent.getPath());
            Log.v("packtchat", "Message received on watch is: " +
            message);

            // Broadcast message to wearable activity for display
            Intent messageIntent = new Intent();
            messageIntent.setAction(Intent.ACTION_SEND);
            messageIntent.putExtra("heart", message);
            LocalBroadcastManager.getInstance(this)
            .sendBroadcast(messageIntent);
        }
        else {
            super.onMessageReceived(messageEvent);
        }
    }
}
```

Register the `service` class in Manifest as follows:

```
<service android:name=".services.HeartListener">
    <intent-filter>
        <action android:name=
        "com.google.android.gms.wearable.DATA_CHANGED" />
        <action android:name=
        "com.google.android.gms.wearable.MESSAGE_RECEIVED" />
```

```
        <data
            android:host="*"
            android:pathPrefix="/heartRate"
            android:scheme="wear" />
    </intent-filter>
</service>
```

In `MainActivity`, we shall write another Broadcast receiver. Let's call it `HeartRateReceiver`:

```
public class HeartRateReciver extends BroadcastReceiver {
    @Override
    public void onReceive(Context context, Intent intent) {
        String data = intent.getStringExtra("heart");
        Log.v("heart", "Main activity received message: " +
        message);

        mHeart.setText(message);
    }
}
```

Register `BroadcastReceiver` in the `oncreate` method as follows:

```
// Register the local broadcast receiver
IntentFilter DataFilter = new IntentFilter(Intent.ACTION_SEND);
HeartRateReciver DataReceiver = new HeartRateReciver();
LocalBroadcastManager.getInstance(this).registerReceiver(DataReceiver,
DataFilter);
```

We are successfully receiving heart rate data directly from `HeartListener` to `broadcastreceiver`. Now, let's work on the User Interface for a Mobile project. We need to keep the UI comprehensively simple and powerful; the following design talks about interoperability with the Wear app and distance and calorie burn prediction.

Conceptualizing the application

The upbeat mobile application should display the steps and pulse rate. Upbeat needs to send a pulse rate request to the Wear app. History shows the distance and calories burned from the database. Reset will clear the database.

Landing screen: when the user opens the app, he or she will see something similar to the following design:

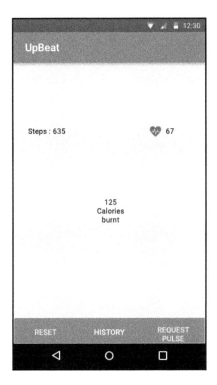

Before we start working on the design, we need to be sure of a couple of things, such as the colors, backgrounds, and so on. Inside the `res/values` directory, open the `colors.xml` file and add the following color values:

```
<?xml version="1.0" encoding="utf-8"?>
<resources>
    <color name="colorPrimary">#607d8b</color>
    <color name="colorPrimaryDark">#34515e</color>
    <color name="colorAccent">#FFF</color>
    <color name="grey">#afaeae</color>
    <color name="white">#fff</color>
</resources>
```

Create a `drawable` resource file, call it `button_bg.xml`, and add the following selector code for the background selection:

```
<?xml version="1.0" encoding="utf-8"?>
<selector xmlns:android="http://schemas.android.com/apk/res/android">
```

```
        <item android:drawable="@color/colorPrimaryDark"
        android:state_pressed="true"/>
        <item android:drawable="@color/grey" android:state_focused="true"/>
        <item android:drawable="@color/colorPrimary"/>
    </selector>
```

In `activity_main.xml`, we need three buttons for the planned functionality as per the design and three textviews. We will use a relative layout as the root container and the following code explains how to do it:

```
<?xml version="1.0" encoding="utf-8"?>
<RelativeLayout xmlns:android="http://schemas.android.com/apk/res/android"
    xmlns:app="http://schemas.android.com/apk/res-auto"
    xmlns:tools="http://schemas.android.com/tools"
    android:layout_width="match_parent"
    android:layout_height="match_parent"
    tools:context="com.packt.upbeat.MainActivity">

    <LinearLayout
        android:orientation="horizontal"
        android:layout_width="match_parent"
        android:layout_height="wrap_content"
        android:layout_marginLeft="10dp"
        android:layout_marginRight="10dp"
        android:layout_marginBottom="102dp"
        android:layout_above="@+id/calory"
        android:layout_centerHorizontal="true">

        <TextView
            android:id="@+id/steps"
            android:layout_width="wrap_content"
            android:layout_height="wrap_content"
            android:layout_alignBottom="@+id/linearLayout"
            android:layout_toStartOf="@+id/calory"
            android:layout_weight="1"
            android:text="Steps!"
            android:textColor="@color/colorPrimaryDark"
            android:textSize="30sp"
            android:textStyle="bold"
            app:layout_constraintBottom_toBottomOf="parent"
            app:layout_constraintLeft_toLeftOf="parent"
            app:layout_constraintRight_toRightOf="parent"
            app:layout_constraintTop_toTopOf="parent" />

        <LinearLayout
            android:layout_weight="1"
            android:id="@+id/linearLayout"
            android:layout_width="wrap_content"
```

```
        android:layout_height="wrap_content"
        android:gravity="center">

        <ImageView
            android:id="@+id/heartbeat"
            android:layout_width="wrap_content"
            android:layout_height="wrap_content"
            android:layout_alignStart="@+id/heart"
            android:layout_below="@+id/heart"/>

        <TextView
            android:id="@+id/heart"
            android:layout_width="wrap_content"
            android:layout_height="wrap_content"
            android:layout_alignBaseline="@+id/steps"
            android:layout_alignBottom="@+id/steps"
            android:layout_alignParentEnd="true"
            android:layout_marginEnd="25dp"
            android:text="Heart!"
            android:textColor="@color/colorPrimaryDark"
            android:textSize="30sp"
            android:textStyle="bold" />
    </LinearLayout>

</LinearLayout>


<TextView
    android:id="@+id/calory"
    android:layout_width="wrap_content"
    android:layout_height="wrap_content"
    android:gravity="center"
    android:text="Calories!"
    android:textColor="@color/colorPrimary"
    android:textSize="20sp"
    android:textStyle="bold"
    android:layout_centerVertical="true"
    android:layout_centerHorizontal="true" />

<LinearLayout
    android:layout_width="match_parent"
    android:layout_height="wrap_content"
    android:layout_alignParentBottom="true"
    android:layout_alignParentStart="true"
    android:orientation="horizontal">

    <android.support.v7.widget.AppCompatButton
        android:id="@+id/reset"
```

```
            android:layout_width="match_parent"
            android:layout_height="50dp"
            android:layout_weight="1"
            android:background="@drawable/button_background"
            android:elevation="5dp"
            android:gravity="center"
            android:text="Reset"
            android:textAllCaps="true"
            android:textColor="@color/white"
            android:textStyle="bold" />

        <android.support.v7.widget.AppCompatButton
            android:id="@+id/history"
            android:layout_width="match_parent"
            android:layout_height="50dp"
            android:layout_weight="1"
            android:background="@drawable/button_background"
            android:elevation="5dp"
            android:gravity="center"
            android:text="History"
            android:textAllCaps="true"
            android:textColor="@color/white"
            android:textStyle="bold" />

        <android.support.v7.widget.AppCompatButton
            android:id="@+id/pulseRequest"
            android:layout_width="match_parent"
            android:layout_height="50dp"
            android:layout_weight="1"
            android:background="@drawable/button_background"
            android:elevation="5dp"
            android:gravity="center"
            android:text="Request pulse"
            android:textAllCaps="true"
            android:textColor="@color/white"
            android:textStyle="bold" />
    </LinearLayout>
</RelativeLayout>
```

To display heart rate, we have `LinearLayour` with `Imageview` and `Textview`, where `imageview` is going to be static. Instead, replace `imageview` with `HeartBeatView` we created in the Wear module for heart custom animation. Let's create it one last time.

Inside the `res/values` folder, add the `heartbeatview_attrs.xml` file and add the following code:

```xml
<?xml version="1.0" encoding="utf-8"?>
<resources>

    <declare-styleable name="HeartBeatView">
        <attr name="scaleFactor" format="float" />
        <attr name="duration" format="integer" />
    </declare-styleable>

</resources>
```

Inside drawables, create a vector graphic XML file and add the following code inside for achieving the heart shape:

```xml
<vector xmlns:android="http://schemas.android.com/apk/res/android"
    android:width="24dp"
    android:height="24dp"
    android:viewportHeight="24.0"
    android:viewportWidth="24.0">
    <path
        android:fillColor="#FFFF0000"
        android:pathData="M12,21.35l-1.45,-1.32C5.4,15.36 2,12.28 2,8.5
        2,5.42 4.42,3 7.5,3c1.74,0 3.41,0.81 4.5,2.09C13.09,3.81
        14.76,3 16.5,3 19.58,3 22,5.42 22,8.5c0,3.78 -3.4,6.86
        -8.55,11.54L12,21.35z"/>
</vector>
```

We can create a class called `HeartBearView` inside the utils package and add the following code for all the animations and custom view logic. For more details on the implementation, you can refer to the Wear module `HeartBeatView` class in the previous chapter:

```java
public class HeartBeatView extends AppCompatImageView {

    private static final String TAG = "HeartBeatView";

    private static final float DEFAULT_SCALE_FACTOR = 0.2f;
    private static final int DEFAULT_DURATION = 50;
    private Drawable heartDrawable;

    private boolean heartBeating = false;

    float scaleFactor = DEFAULT_SCALE_FACTOR;
    float reductionScaleFactor = -scaleFactor;
    int duration = DEFAULT_DURATION;

    public HeartBeatView(Context context) {
```

```
        super(context);
        init();
    }

    public HeartBeatView(Context context, AttributeSet attrs) {
        super(context, attrs);
        populateFromAttributes(context, attrs);
        init();
    }

    public HeartBeatView(Context context, AttributeSet attrs, int
defStyleAttr) {
        super(context, attrs, defStyleAttr);
        populateFromAttributes(context, attrs);
        init();
    }

    private void init() {
        //make this not mandatory
        heartDrawable = ContextCompat.getDrawable(getContext(),
        R.drawable.ic_heart_red_24dp);
        setImageDrawable(heartDrawable);

    }

    private void populateFromAttributes(Context context, AttributeSet
attrs) {
        TypedArray a = context.getTheme().obtainStyledAttributes(
                attrs,
                R.styleable.HeartBeatView,
                0, 0
        );
        try {
            scaleFactor = a.getFloat(R.styleable
            .HeartBeatView_scaleFactor, DEFAULT_SCALE_FACTOR);
            reductionScaleFactor = -scaleFactor;
            duration = a.getInteger(R.styleable.HeartBeatView_duration,
            DEFAULT_DURATION);


        } finally {
            a.recycle();
        }

    }


    /**
```

```
 * toggles current heat beat state
 */
public void toggle() {
    if (heartBeating) {
        stop();
    } else {
        start();
    }
}

/**
 * Starts the heat beat/pump animation
 */
public void start() {
    heartBeating = true;
    animate().scaleXBy(scaleFactor)
    .scaleYBy(scaleFactor).setDuration(duration)
    .setListener(scaleUpListener);
}

/**
 * Stops the heat beat/pump animation
 */
public void stop() {
    heartBeating = false;
    clearAnimation();
}

/**
 * is the heart currently beating
 *
 * @return
 */
public boolean isHeartBeating() {
    return heartBeating;
}

public int getDuration() {
    return duration;
}


private static final int milliInMinute = 60000;

/**
 * set the duration of the beat based on the beats per minute
 *
 * @param bpm (positive int above 0)
```

```
  */
public void setDurationBasedOnBPM(int bpm) {
    if (bpm > 0) {
        duration = Math.round((milliInMinute / bpm) / 3f);
    }
}


public void setDuration(int duration) {
    this.duration = duration;
}

public float getScaleFactor() {
    return scaleFactor;
}

public void setScaleFactor(float scaleFactor) {
    this.scaleFactor = scaleFactor;
    reductionScaleFactor = -scaleFactor;
}

private final Animator.AnimatorListener scaleUpListener = new
Animator.AnimatorListener() {

    @Override
    public void onAnimationStart(Animator animation) {
    }

    @Override
    public void onAnimationRepeat(Animator animation) {

    }

    @Override
    public void onAnimationEnd(Animator animation) {
        //we ignore heartBeating as we want to ensure the heart is
        reduced back to original size
        animate().scaleXBy(reductionScaleFactor)
        .scaleYBy(reductionScaleFactor).setDuration(duration)
        .setListener(scaleDownListener);
    }

    @Override
    public void onAnimationCancel(Animator animation) {

    }
};
```

```
private final Animator.AnimatorListener scaleDownListener = new
Animator.AnimatorListener() {

    @Override
    public void onAnimationStart(Animator animation) {
    }

    @Override
    public void onAnimationRepeat(Animator animation) {
    }

    @Override
    public void onAnimationEnd(Animator animation) {
        if (heartBeating) {
            //duration twice as long for the upscale
            animate().scaleXBy(scaleFactor).scaleYBy(scaleFactor)
            .setDuration
            (duration * 2).setListener(scaleUpListener);
        }
    }

    @Override
    public void onAnimationCancel(Animator animation) {
    }
};
}
```

Inside the `activity_main.xml` file, instead of `ImageView`, replace the code with a custom view created in the mobile project scope:

```
<com.packt.upbeat.utils.HeartBeatView
    android:id="@+id/heartbeat"
    android:layout_width="wrap_content"
    android:layout_height="wrap_content"
    android:layout_alignStart="@+id/heart"
    android:layout_below="@+id/heart"/>
```

Now that our landing page user interface is ready, we can get started by working on `MainActivity`.

In `MainActivity`, let's instantiate all the UI components we used in the layout:

```
private AppCompatButton mReset, mHistory, mHeartPulse;
private TextView mSteps, mHeart, mCalory;
private HeartBeatView heartbeat;
```

Map the components with its ID using the `findviewbyid` method inside the `oncreate` method:

```
heartbeat = (HeartBeatView) findViewById(R.id.heartbeat);
mSteps = (TextView) findViewById(R.id.steps);
mHeart = (TextView) findViewById(R.id.heart);
mCalory = (TextView) findViewById(R.id.calory);
mReset = (AppCompatButton) findViewById(R.id.reset);
mHistory = (AppCompatButton) findViewById(R.id.history);
mHeartPulse = (AppCompatButton) findViewById(R.id.pulseRequest);
```

Inside the `HeartRateReceiver` class, fetch the data, convert the data into integers, and show it in the UI. The following code illustrates activating `HeartBeatAnimation` with data received from the Wear app:

```
@Override
    public void onReceive(Context context, Intent intent) {
        String data = intent.getStringExtra("heart");
        Log.v("heart", "Main activity received message: " + data);

        mHeart.setText(data);
        heartbeat.setDurationBasedOnBPM(Integer.valueOf(data));
        heartbeat.toggle();
    }
```

In `StepReceiver`, we shall set the data to the step count `textview` labelled as `mSteps`:

```
mSteps.setText("Steps:"+ message);
```

We have completed receiving the pulse count and step count and showing it in the phone's UI. Now, we need to show the calories burned for those steps.

Finding the calories burned from the step count can be achieved by a number of different methods based on your body mass index and so on. Research on this pedometer steps to calories have introduced a conversion factor, which is as follows:

Conversion factor = 99.75 calories per mile / 2,200 steps per mile = 0.045 calories per step

So using this value, we can determine the calories by simply multiplying this value with the steps count.

Calories burned = 7,000 steps x 0.045 calories per step = 318 calories

Within the `StepReceiver` class, inside the `onReceive` method, add the following code:

```
int value = Integer.valueOf(message);
mCalory.setText(String.valueOf((int)(value * 0.045)) + "\ncalories" +
"\nburnt");
```

The calorie burn and pulse rate checking from the phone is completed. We still have more work in `MainActivity`. We need to persist the data of the step counter to show the history of calories and distance. Let's use `RealmDB`, which we tried in the first chapter.

Add the following classpath to the project level gradle file:

```
classpath "io.realm:realm-gradle-plugin:2.2.1"
```

Apply the previous plugin in the gradle mobile module:

```
apply plugin: 'realm-android'
```

Realm is ready in the project. Now, we need setters and getters for the step data. Add the following class to the models package:

```
public class StepCounts extends RealmObject {

    private String ReceivedDateTime;
    private String Data;

    public String getReceivedDateTime() {
        return ReceivedDateTime;
    }

    public void setReceivedDateTime(String receivedDateTime) {
        ReceivedDateTime = receivedDateTime;
    }

    public String getData() {
        return Data;
    }

    public void setData(String data) {
        Data = data;
    }
}
```

In `MainActivity`, instantiate the Realm and initialize it in the `onCreate` method, as follows:

```
private Realm realm;
```

```
@Override
protected void onCreate(Bundle savedInstanceState) {
...
Realm.init(this);
realm = Realm.getDefaultInstance();

}
```

When the step count is received, add the data to `RealmDB`. Add the following code inside the `onRecieve` method of the `StepReciever` inner class:

```
realm.beginTransaction();
StepCounts Steps = realm.createObject(StepCounts.class);
Steps.setData(message);
String TimeStamp =
DateFormat.getDateTimeInstance().format(System.currentTimeMillis());
Steps.setReceivedDateTime(TimeStamp);
realm.commitTransaction();
```

To show the last value in the UI, add the following code to the `onCreate` method:

```
RealmResults<StepCounts> results = realm.where(StepCounts.class).findAll();

if(results.size() == 0){
    mSteps.setText("Steps: ");
}else{
    mSteps.setText("Steps: "+results.get(results.size()-1).getData());
    int value = Integer.valueOf(results
    .get(results.size()-1).getData());
    mCalory.setText(String.valueOf((int)(value * 0.045))
    + "\ncalories" + "\nburnt");
}
```

For the buttons, now attach the click listeners to the `oncreate` method:

```
mHistory.setOnClickListener(new View.OnClickListener() {
    @Override
    public void onClick(View v) {
    }
});

mHeartPulse.setOnClickListener(new View.OnClickListener() {
    @Override
    public void onClick(View v) {
    }
});

mReset.setOnClickListener(new View.OnClickListener() {
```

```
    @Override
    public void onClick(View v) {
    }
});
```

Let's create another Activity and call it `HistoryActivity`, which will show the list of data received. In the `activity_history.xml` file, add the following code:

```xml
<?xml version="1.0" encoding="utf-8"?>
<LinearLayout xmlns:android="http://schemas.android.com/apk/res/android"
    xmlns:app="http://schemas.android.com/apk/res-auto"
    xmlns:tools="http://schemas.android.com/tools"
    android:layout_width="match_parent"
    android:layout_height="match_parent"
    tools:context="com.packt.upbeat.HistoryActivity">

    <android.support.v7.widget.RecyclerView
        android:id="@+id/recycler_view"
        android:layout_width="match_parent"
        android:layout_height="match_parent"
        android:layout_margin="5dp" />

</LinearLayout>
```

Now, we need `row_layout` for each item in `recyclerview` and the layout is shown as follows:

```xml
<?xml version="1.0" encoding="utf-8"?>
<LinearLayout xmlns:android="http://schemas.android.com/apk/res/android"
    xmlns:card_view="http://schemas.android.com/apk/res-auto"
    android:layout_width="match_parent"
    android:layout_height="wrap_content"
    android:orientation="vertical">

    <android.support.v7.widget.CardView
        android:layout_width="match_parent"
        android:layout_height="wrap_content"
        android:layout_marginBottom="0dp"
        android:layout_marginLeft="5dp"
        android:layout_marginRight="5dp"
        android:layout_marginTop="9dp"
        card_view:cardCornerRadius="3dp"
        card_view:cardElevation="0.01dp">

        <LinearLayout
            android:layout_margin="10dp"
            android:orientation="vertical"
            android:id="@+id/top_layout"
```

```
        android:layout_width="match_parent"
        android:layout_height="wrap_content">

    <TextView
        android:layout_margin="10dp"
        android:id="@+id/steps"
        android:text="Steps"
        android:layout_width="match_parent"
        android:layout_height="wrap_content" />
    <TextView
        android:layout_margin="10dp"
        android:id="@+id/calories"
        android:text="calory"
        android:layout_width="match_parent"
        android:layout_height="wrap_content" />
    <TextView
        android:layout_margin="10dp"
        android:id="@+id/distance"
        android:text="distance"
        android:layout_width="match_parent"
        android:layout_height="wrap_content" />


    <TextView
        android:layout_margin="10dp"
        android:id="@+id/date"
        android:layout_width="match_parent"
        android:layout_height="40dp"
        android:layout_gravity="bottom"
        android:background="#ff444444"
        android:gravity="center"
        android:text="Timestamp"
        android:textColor="#fff"
        android:textSize="20dp" />

    </LinearLayout>
    </android.support.v7.widget.CardView>
</LinearLayout>
```

 Remember, before using `cardview` and `recyclerview`, we need to add the support dependencies to our gradle module:
```
compile 'com.android.support:cardview-v7:25.1.1'
compile 'com.android.support:recyclerview-v7:25.1.1'
```

Recyclerview Adapter

We will have to create an `adapter` class that fetches data from the Realm and adapts to `row_layout` that is created:

```
public class HistoryAdapter extends
RecyclerView.Adapter<HistoryAdapter.ViewHolder> {

    public List<StepCounts> steps;
    public Context mContext;

    public HistoryAdapter(List<StepCounts> steps, Context mContext) {
        this.steps = steps;
        this.mContext = mContext;
    }

    @Override
    public ViewHolder onCreateViewHolder(ViewGroup viewGroup, int i) {
        View v = LayoutInflater.from(viewGroup.getContext())
                .inflate(R.layout.row_item, viewGroup, false);
        ViewHolder viewHolder = new ViewHolder(v);
        return viewHolder;
    }

    @Override
    public void onBindViewHolder(ViewHolder viewHolder, int i) {
        viewHolder.steps.setText(steps.get(i).getData()+" Steps");
        viewHolder.date.setText(steps.get(i).getReceivedDateTime());

        int value = Integer.valueOf(steps.get(i).getData());
        DecimalFormat df = new DecimalFormat("#.00") ;
        String kms = String.valueOf(df.format(value * 0.000762)) + "
        kms" ;
        viewHolder.calory.setText(String.valueOf((int)(value * 0.045))
        + " calories " + "burnt");
        viewHolder.distance.setText("Distance: "+kms);

    }

    @Override
    public int getItemCount() {
        return steps.size();
    }


    public static class ViewHolder extends RecyclerView.ViewHolder  {

        public TextView steps,calory,distance,date;
```

```
      public ViewHolder(View itemView) {
          super(itemView);
          steps = (TextView) itemView.findViewById(R.id.steps);
          calory = (TextView) itemView.findViewById(R.id.calories);
          distance = (TextView) itemView.findViewById(R.id.distance);
          date = (TextView) itemView.findViewById(R.id.date);
      }
  }

}
```

In adapter, we are showing calories burned with the conversion factor value. For finding a generic distance, we have another value and we need to multiply the steps to it, as shown in the adapter.

In `HistoryActivity`, in the class global scope, declare the following instances:

```
Realm realm;
RecyclerView mRecyclerView;
RecyclerView.LayoutManager mLayoutManager;
RecyclerView.Adapter mAdapter;
```

Now, in the `oncreate` method of the `HistoryActivity` class, add the following code:

```
mRecyclerView = (RecyclerView) findViewById(R.id.recycler_view);
mRecyclerView.setHasFixedSize(true);

Realm.init(this);
realm = Realm.getDefaultInstance();
RealmResults<StepCounts> results = realm.where(StepCounts.class).findAll();
// The number of Columns
mLayoutManager = new GridLayoutManager(this, 1);
mRecyclerView.setLayoutManager(mLayoutManager);

mAdapter = new HistoryAdapter(results,HistoryActivity.this);
mRecyclerView.setAdapter(mAdapter);
```

Completed HistoryActivity Class

The complete class would look as follows:

```
public class HistoryActivity extends AppCompatActivity {

    Realm realm;
    RecyclerView mRecyclerView;
    RecyclerView.LayoutManager mLayoutManager;
    RecyclerView.Adapter mAdapter;
```

```
@Override
protected void onCreate(Bundle savedInstanceState) {
    super.onCreate(savedInstanceState);
    setContentView(R.layout.activity_history);
    // Calling the RecyclerView
    mRecyclerView = (RecyclerView)
    findViewById(R.id.recycler_view);
    mRecyclerView.setHasFixedSize(true);

    Realm.init(this);
    realm = Realm.getDefaultInstance();
    RealmResults<StepCounts> results =
    realm.where(StepCounts.class).findAll();
    // The number of Columns
    mLayoutManager = new GridLayoutManager(this, 1);
    mRecyclerView.setLayoutManager(mLayoutManager);

    mAdapter = new HistoryAdapter(results,HistoryActivity.this);
    mRecyclerView.setAdapter(mAdapter);
    }
}
```

In `MainActivity`, **start** `historyActivity` when the `mHistory` button is clicked:

```
mHistory.setOnClickListener(new View.OnClickListener() {
    @Override
    public void onClick(View v) {
        startActivity(new Intent(MainActivity.this,
        HistoryActivity.class));
    }
});
```

Now, it's time to send the data from the mobile to wear using the same method that we used in Wear.

We will create a class that extends `Thread` and, using Node and Message API, we will send the data as follows:

```
class SendToDataLayerThread extends Thread {
    String path;
    String message;

    // Constructor to send a message to the data layer
    SendToDataLayerThread(String p, String msg) {
        path = p;
        message = msg;
    }
```

```
    public void run() {
        NodeApi.GetConnectedNodesResult nodes =
        Wearable.NodeApi.getConnectedNodes(googleClient).await();
        for (Node node : nodes.getNodes()) {
            MessageApi.SendMessageResult result =
            Wearable.MessageApi.sendMessage(googleClient,
            node.getId(), path, message.getBytes()).await();
            if (result.getStatus().isSuccess()) {
                Log.v("myTag", "Message: {" + message + "} sent to: " +
                node.getDisplayName());
            } else {
                // Log an error
                Log.v("myTag", "ERROR: failed to send Message");
            }
        }
    }
}
```

Inside the `mHeartPulse` button click listener, and start the `SendToDataLayerThread` class as follows:

```
mHeartPulse.setOnClickListener(new View.OnClickListener() {
    @Override
    public void onClick(View v) {
        new SendToDataLayerThread("/heart", "Start upbeat for heart
        rate").start();
    }
});
```

Now, switch back to the Wear project scope and add a new class that extends to `WearableListenerService`. When it receives a message from the mobile app, then fire a notification to start the application. The complete class code is as follows:

```
public class MobileListener extends WearableListenerService {

    @Override
    public void onMessageReceived(MessageEvent messageEvent) {

        if (messageEvent.getPath().equals("/heart")) {
            final String message = new String(messageEvent.getData());
            Log.v("myTag", "Message path received on watch is: " +
            messageEvent.getPath());
            Log.v("myTag", "Message received on watch is: " + message);

            // Broadcast message to wearable activity for display
            Intent messageIntent = new Intent();
            messageIntent.setAction(Intent.ACTION_SEND);
            messageIntent.putExtra("message", message);
```

```
LocalBroadcastManager.getInstance(this)
.sendBroadcast(messageIntent);

Intent intent2 = new Intent
(getApplicationContext(), MainActivity.class);

PendingIntent pendingIntent = PendingIntent.getActivity
(getApplicationContext(), 0, intent2,
        PendingIntent.FLAG_ONE_SHOT);

Uri defaultSoundUri = RingtoneManager.getDefaultUri
(RingtoneManager.TYPE_ALARM);

NotificationCompat.Builder notificationBuilder =
(NotificationCompat.Builder) new
NotificationCompat.Builder(getApplicationContext())
        .setAutoCancel(true)    //Automatically delete the
        notification
        .setSmallIcon(R.drawable.ic_heart_icon)
        //Notification icon
        .setContentIntent(pendingIntent)
        .setContentTitle("Open upbeat")
        .setContentText("UpBeat to check the pulse")
        .setCategory(Notification.CATEGORY_REMINDER)
        .setPriority(Notification.PRIORITY_HIGH)
        .setSound(defaultSoundUri);


NotificationManagerCompat notificationManager =
NotificationManagerCompat.from
(getApplicationContext());
notificationManager.notify(0, notificationBuilder.build());

    }
    else {
        super.onMessageReceived(messageEvent);
    }
  }
}
```

Now, register the previously mentioned service in the manifest with the correct path mobile app using the following:

```
<service android:name=".services.MobileListener">
    <intent-filter>
        <action android:name=
        "com.google.android.gms.wearable.DATA_CHANGED" />
        <action android:name=
        "com.google.android.gms.wearable.MESSAGE_RECEIVED" />

        <data
            android:host="*"
            android:pathPrefix="/heart"
            android:scheme="wear" />
    </intent-filter>
</service>
```

Let's switch back to the mobile project scope and finish the reset button click event. We will write a method that flushes the `RealmDB` data and recreates the activity:

```
public void Reset(){
    RealmResults<StepCounts> results =
    realm.where(StepCounts.class).findAll();

    realm.beginTransaction();

    results.deleteAllFromRealm();

    realm.commitTransaction();
}
```

Inside the click listener, add the following methods as follows:

```
mReset.setOnClickListener(new View.OnClickListener() {
    @Override
    public void onClick(View v) {
        Reset();
        recreate();
    }
});
```

Switch to Wear project scope and create a new Activity for health tips, and we will call the activity `HealthTipsActivity`. Here, on this screen, we will list a few good health tips and suggestions.

In `activity_health_tips.xml`, add the following code:

```
<?xml version="1.0" encoding="utf-8"?>
<android.support.wearable.view.BoxInsetLayout
xmlns:android="http://schemas.android.com/apk/res/android"
    xmlns:app="http://schemas.android.com/apk/res-auto"
    xmlns:tools="http://schemas.android.com/tools"
    android:id="@+id/container"
    android:layout_width="match_parent"
    android:layout_height="match_parent"
    android:padding="5dp"
    app:layout_box="all"
    tools:deviceIds="wear">

    <android.support.wearable.view.WearableRecyclerView
        android:id="@+id/wearable_recycler_view"
        android:layout_width="match_parent"
        android:layout_height="match_parent" />


</android.support.wearable.view.BoxInsetLayout>
```

We need to add one more layout for the row item for the tips activity. We will call this layout `health_tips_row.xml`:

```
<?xml version="1.0" encoding="utf-8"?>
<LinearLayout xmlns:android="http://schemas.android.com/apk/res/android"
    android:layout_width="match_parent"
    android:layout_height="match_parent"
    android:orientation="vertical"
    android:tag="cards main container">

    <android.support.v7.widget.CardView
    xmlns:card_view="http://schemas.android.com/apk/res-auto"
        android:id="@+id/card_view"
        android:layout_width="match_parent"
        android:layout_height="wrap_content"
        card_view:cardBackgroundColor="@color/colorPrimary"
        card_view:cardCornerRadius="10dp"
        card_view:cardElevation="5dp"
        card_view:cardUseCompatPadding="true">

            <LinearLayout
                android:layout_width="match_parent"
                android:layout_height="wrap_content"
                android:layout_marginTop="12dp"
                android:layout_weight="2"
                android:orientation="vertical">
```

```xml
                <TextView
                    android:id="@+id/health_tip"
                    android:layout_width="wrap_content"
                    android:layout_height="wrap_content"
                    android:layout_gravity="center_horizontal"
                    android:layout_marginTop="10dp"
                    android:text="HealthTip"
                    android:textColor="@color/white"
                    android:textAppearance="?
                    android:attr/textAppearanceLarge" />

                <TextView
                    android:id="@+id/tip_details"
                    android:layout_width="wrap_content"
                    android:layout_height="wrap_content"
                    android:layout_gravity="center_horizontal"
                    android:layout_marginTop="10dp"
                    android:text="Details"
                    android:textColor="@color/white"
                    android:textAppearance="?
                    android:attr/textAppearanceMedium" />

            </LinearLayout>

        </android.support.v7.widget.CardView>

    </LinearLayout>
```

Create a model that contains the required fields. We shall create setters and getters with a full parameterized constructor for all the fields:

```java
public class HealthTipsItem {

    public String Title;
    public String MoreInfo;

    public HealthTipsItem(String title, String moreInfo) {
        Title = title;
        MoreInfo = moreInfo;
    }

    public String getTitle() {
        return Title;
    }

    public void setTitle(String title) {
        Title = title;
    }
```

```
    public String getMoreInfo() {
        return MoreInfo;
    }

    public void setMoreInfo(String moreInfo) {
        MoreInfo = moreInfo;
    }
}
```

We shall have another data class that will keep all the health tips:

```
public class HealthTips {

    public static String[] nameArray =
            {"Food style",
                "Food style",
                "Food style",
                "Drinking water",
                "Unhealthy drinks",
                "Alcohol and drugs",
                "Body Mass index",
                "Physical excercise",
                "Physical activities",
                "Meditation",
                "Healthy signs"};

    public static String[] versionArray = {
                "Along with fresh vegetables and fruits, eat lean meats (if
                you're not vegetarian), nuts, and seeds.",
                "Opt for seasonal and local products instead of those
                exotic imported foodstuff",
                "Make sure you get a proper balanced diet, as often as
                possible",
                "Drink water - you need to stay hydrated. It is great for
                your internal organs, and it also keeps your skin healthy
                and diminishes acne",
                "Stop drinking too much caffeine and caffeinated
                beverages",
                "Limit alcohol intake. Tobacco and drugs should be a firm
                No",
                "Maintain a healthy weight.",
                "Exercise at least four days a week for 20 to 30 minutes
                each day. Another option is to break your workouts into
                several sessions",
                "Try to have as much physical activity as you can. Take the
                 stairs instead of elevator; walk to the market instead of
                 taking your car etc",
                "Practice simple meditation. It balances your body, mind,
```

```
                    and soul",
                    "When speaking about health tips, skin, teeth, hair, and
                    nails are all health signs. Loss of hair or fragile nails
                    might mean poor nutrition"};

     }
```

Now, we shall create an adapter to work with the list of health advice. The following code takes the data and loads it in `wearablerecyclerview`:

```java
public class RecyclerViewAdapter
        extends WearableRecyclerView.Adapter
        <RecyclerViewAdapter.ViewHolder> {

    private List<HealthTipsItem> mListTips = new ArrayList<>();
    private Context mContext;

    public RecyclerViewAdapter(List<HealthTipsItem> mListTips, Context
    mContext) {
        this.mListTips = mListTips;
        this.mContext = mContext;
    }

    static class ViewHolder extends RecyclerView.ViewHolder {
        private TextView Title, info;

        ViewHolder(View view) {
            super(view);
            Title = (TextView) view.findViewById(R.id.health_tip);
            info = (TextView) view.findViewById(R.id.tip_details);
        }
    }

    @Override
    public RecyclerViewAdapter.ViewHolder onCreateViewHolder(ViewGroup
    parent, int viewType) {
        View view = LayoutInflater.from(parent.getContext())
                .inflate(R.layout.health_tips_row, parent, false);

        return new ViewHolder(view);
    }

    @Override
    public void onBindViewHolder(ViewHolder holder, int position) {

        holder.Title.setText(mListTips.get(position).getTitle());
        holder.info.setText(mListTips.get(position).getMoreInfo());
    }
```

```
    @Override
    public int getItemCount() {
        return mListTips.size();
    }
}
```

Within the global scope of the activity, declare the following instances:

```
private RecyclerViewAdapter mAdapter;
private List<HealthTipsItem> myDataSet = new ArrayList<>();
```

Inside the `oncreate` method, we can complete the app by adding the following code:

```
WearableRecyclerView recyclerView = (WearableRecyclerView)
findViewById(R.id.wearable_recycler_view);
recyclerView.setHasFixedSize(true);
LinearLayoutManager mLayoutManager = new LinearLayoutManager(this);
recyclerView.setLayoutManager(mLayoutManager);

myDataSet = new ArrayList<HealthTipsItem>();
for (int i = 0; i < HealthTips.nameArray.length; i++) {
    myDataSet.add(new HealthTipsItem(
            HealthTips.nameArray[i],
            HealthTips.versionArray[i]
    ));
}

mAdapter = new RecyclerViewAdapter(myDataSet,HealthTipsActivity.this);
recyclerView.setAdapter(mAdapter);
```

Let's create another Activity for a generic calories chart from an international food list and call the activity `CalorychartActivity`.

In the `CaloryChartActivity` layout file, we will add the `WearableRecyclerView` component:

```
<?xml version="1.0" encoding="utf-8"?>
<android.support.wearable.view.BoxInsetLayout
xmlns:android="http://schemas.android.com/apk/res/android"
    xmlns:app="http://schemas.android.com/apk/res-auto"
    xmlns:tools="http://schemas.android.com/tools"
    android:id="@+id/container"
    android:layout_width="match_parent"
    android:layout_height="match_parent"
    android:padding="5dp"
    app:layout_box="all"
    tools:deviceIds="wear">
```

```xml
    <android.support.wearable.view.WearableRecyclerView
        android:id="@+id/wearable_recycler_view"
        android:layout_width="match_parent"
        android:layout_height="match_parent" />


</android.support.wearable.view.BoxInsetLayout>
```

Create another layout for each calorie chart item and add the following code:

```xml
<?xml version="1.0" encoding="utf-8"?>
<LinearLayout xmlns:android="http://schemas.android.com/apk/res/android"
    android:layout_width="match_parent"
    android:layout_height="match_parent"
    android:orientation="vertical"
    android:tag="cards main container">

    <android.support.v7.widget.CardView
    xmlns:card_view="http://schemas.android.com/apk/res-auto"
        android:id="@+id/card_view"
        android:layout_width="match_parent"
        android:layout_height="wrap_content"
        card_view:cardBackgroundColor="@color/colorPrimary"
        card_view:cardCornerRadius="10dp"
        card_view:cardElevation="5dp"
        card_view:cardUseCompatPadding="true">

            <LinearLayout
                android:layout_width="match_parent"
                android:layout_height="wrap_content"
                android:layout_margin="12dp"
                android:layout_weight="2"
                android:orientation="vertical">

            <TextView
                android:id="@+id/health_tip"
                android:layout_width="wrap_content"
                android:layout_height="wrap_content"
                android:layout_gravity="center_horizontal"
                android:layout_marginTop="10dp"
                android:text="calory"
                android:textColor="@color/white"
                android:textAppearance="?
                android:attr/textAppearanceLarge" />


        </LinearLayout>
```

```
    </android.support.v7.widget.CardView>

</LinearLayout>
```

We shall create the `model` class for the calories as follows:

```
public class CaloryItem {

    public String Calories;

    public CaloryItem(String calories) {
        Calories = calories;
    }

    public String getCalories() {
        return Calories;
    }

    public void setCalories(String calories) {
        Calories = calories;
    }

}
```

We shall create another adapter for the calorie chart. The adapter is similar to the `HealthTips` adapter. Create a file `RecyclerViewCaloryAdapter` and add the following code to it:

```
public class RecyclerViewCaloryAdapter
        extends
WearableRecyclerView.Adapter<RecyclerViewCaloryAdapter.ViewHolder> {

    private List<CaloryItem> mCalory = new ArrayList<>();
    private Context mContext;

    public RecyclerViewCaloryAdapter(List<CaloryItem> mCalory, Context
    mContext) {
        this.mCalory = mCalory;
        this.mContext = mContext;
    }

    static class ViewHolder extends RecyclerView.ViewHolder {
        private TextView Title;

        ViewHolder(View view) {
            super(view);
            Title = (TextView) view.findViewById(R.id.health_tip);
        }
    }
```

```
    @Override
    public RecyclerViewCaloryAdapter.ViewHolder
    onCreateViewHolder(ViewGroup parent, int viewType) {
        View view = LayoutInflater.from(parent.getContext())
                .inflate(R.layout.calory_row, parent, false);

        return new ViewHolder(view);
    }

    @Override
    public void onBindViewHolder(ViewHolder holder, int position) {
        holder.Title.setText(mCalory.get(position).getCalories());
    }

    @Override
    public int getItemCount() {
        return mCalory.size();
    }
}
```

In the `CaloryChartActivity` global scope of the project, add the following instances:

```
private RecyclerViewCaloryAdapter mAdapter;
private List<CaloryItem> myDataSet = new ArrayList<>();
```

Add the following code within the `oncreate` method:

```
WearableRecyclerView recyclerView = (WearableRecyclerView)
findViewById(R.id.wearable_recycler_view);
recyclerView.setHasFixedSize(true);
LinearLayoutManager mLayoutManager = new LinearLayoutManager(this);
recyclerView.setLayoutManager(mLayoutManager);

myDataSet = new ArrayList<CaloryItem>();
for (int i = 0; i < Calory.nameArray.length; i++) {
    myDataSet.add(new CaloryItem(
            Calory.nameArray[i]
    ));
}

mAdapter = new
RecyclerViewCaloryAdapter(myDataSet,CaloryChartActivity.this);
recyclerView.setAdapter(mAdapter);
```

The following screens show the complete mobile and Wear app.

The following image illustrates the active screen that listens to the steps and pulse:

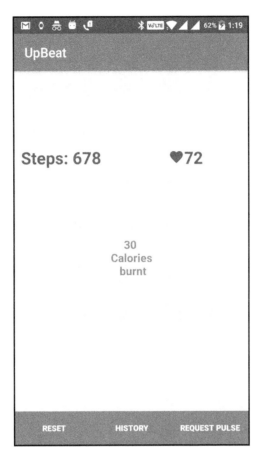

The following image illustrates the healthy food tips screen in the Wear app. It is designed with `WearableRecyclerView`:

The calorie chart for learning the calories available in different food items is shown in the following image:

Summary

In this chapter, we understood the fundamentals of working with Wear and mobile applications. We have explored the API support for sending and receiving data from Wear to mobile and vice versa. Now, integrating `RealmDB` for any Wear projects will be easier.

In the next chapter, we will build a Google Maps application for Wear devices and we will persist the location data, understanding different map types and controls for Wear devices.

6
Ways to Get Around Anywhere - WearMap and the GoogleAPIclient

A Map is a visual representation of an area or a part of an area.

We humans travel to different cities; they could be domestic or international cities. How about tracking the places you visit? We all use maps for different reasons, but in most cases, we use maps to plan a particular activity, such as outdoor tours, cycling, and other similar activities. Maps influence human intelligence to find the fastest route from the source location to the destination. In this project, we will build a Wear application that works with the Google Maps service.

For record, Google Maps started as a C++ desktop program in October, 2004. Google Maps officially released in February, 2005. Google Maps offers an API that allows maps to be embedded in third-party applications; Google Maps offers aerial and satellite views of many places. Google Maps is the best compared to other map services; maps are optimize and its accuracy rate is very good.

In this project, let's build a standalone Wear map application. When a user clicks on the map, we shall allow the user to write a story about the location and save it in the SQLite database as a marker. When a user clicks on the marker, we should show the user what is saved. In this chapter, we will understand the following important concepts:

- Creating a project in the developer API console
- Getting the Maps API key with the help of the SHA1 fingerprint
- SQLite integration
- Google Maps

- The Google API Client and more
- GeoCoder

Let's get started with creating WearMap

We now know how to create a standalone application. In case if you are following this project directly without following Wear-note application which is covered in Chapter 2, *Let us Help Capture What is on Your Mind - WearRecyclerView and More* and Chapter 3, *Let us Help Capture What is on Your Mind - Saving Data and Customizing the UI*. Please do follow the Wear-note application to learn more about standalone applications.

Let's call this project **WearMapDiary**, since we store the locations and details about the location. The project package address up to the developer; in this project, the package address is com.packt.wearmapdiary and the API level 25 Nougat. In the activity template, select the **Google Maps Wear Activity**, as follows:

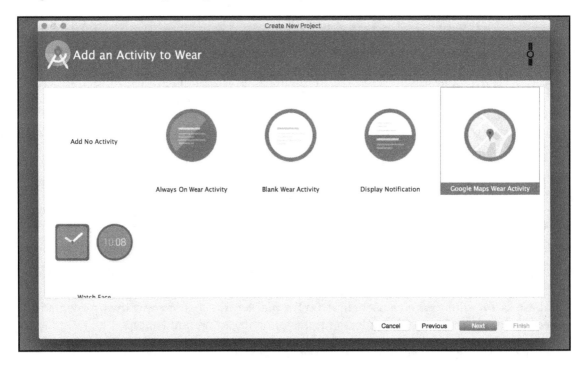

Select Google Maps Wear Activity template from the activity chooser

Once the project is created, we will see the necessary configuration for the project, which includes the map fragment already being added; it would have set the `DismissOverlays` component:

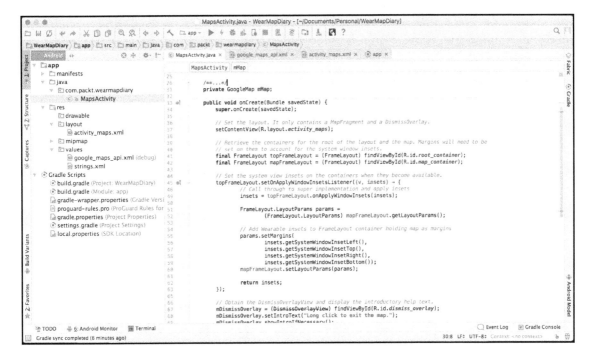

The sample code configured for working with the Wear map activity will be generated.

We need to add the Maps API key to the project in the `res/values` directory `google_maps_api.xml` file:

The Google API console

The Google API console is a web portal that allows developers to manage Google services for their project development and it can be accessed at `https://console.developers.google.com`.

1. Visit the developer console with your Google account. Create a project `packt-wear` or something that is convenient for developers:

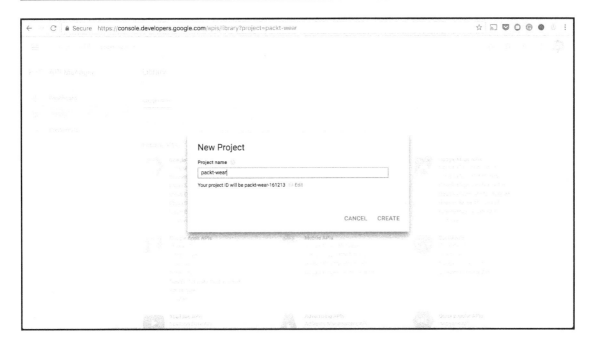

2. After creating the project successfully, go to the **API Manager** | **Library** section and enable the **Google Maps Android API**:

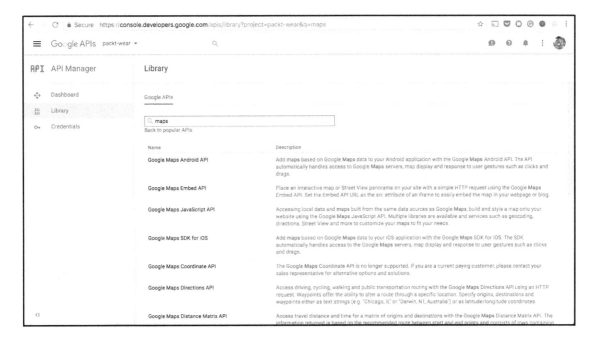

3. Click on the **Enable** button for enabling Maps for Android:

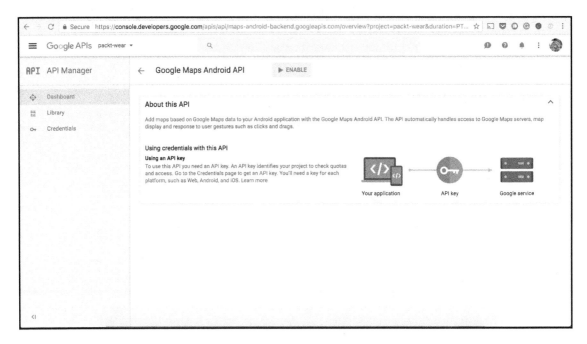

4. After enabling the API in the console facility, we need to create the API key with the development machine's **SHA1** fingerprint and the project's package address, as follows:

5. To get your machine's **SHA1** fingerprint, open Android Studio. On the right-hand side of Android Studio, you will see the Gradle project menu. Then, follow these steps:

    1. Click on **Gradle** (on the right-hand side panel, you will see the **Gradle** Bar)

    2. Click on **Refresh** (click on **Refresh**; on the **Gradle** Bar, you will see a List of Gradle scripts for your project)

    3. Click on **Your Project** (your Project Name from **List** (root))

    4. Click on **Tasks**

    5. Click on **Android**

6. Double-click on **signingReport** (you will get **SHA1** and **MD5** in **Run** Bar):

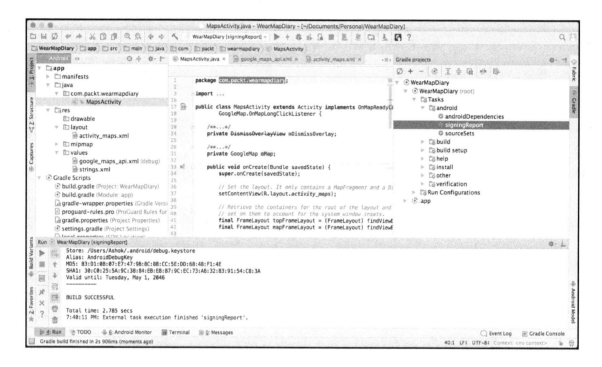

7. Copy your **SHA1** fingerprint, paste it in the Google API console, and save:

8. Now, copy the API key from the console and paste it in the project's `google_maps_api.xml` file, as follows:

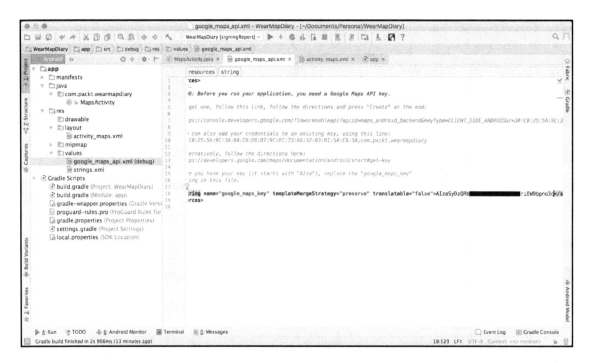

9. Now, switch your Gradle scope to app and compile the project in the Wear emulator or your Wear device:

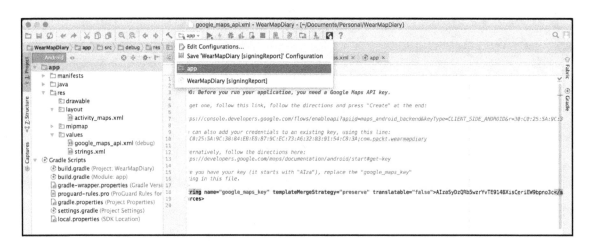

If your Google Play services is not updated in your emulator, Wear throws an error screen to update Play services:

If you have the actual Wear device, the Wear operating system will take care of downloading the latest Google Play services update when its available. For the emulator, we need to connect it to the actual device to add the account. First, connect the Android phone through **adb** and start the Wear emulator.

Install the Android Wear companion app from the Play store `https://play.google.com/store/apps/details?id=com.google.android.wearable.app&hl=en`.

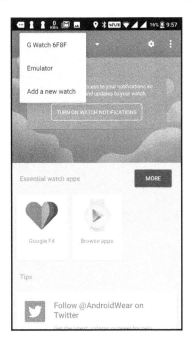

In the android Wear application, select emulator, and in the Android Studio terminal, enter this command:

```
adb -d forward tcp:5601 tcp:5601
```

After the emulator connects to your actual phone, you can add the account that is synced to your phone already or you can add a new account.

The following image illustrates the Wear account's sync screen:

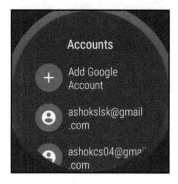

- After successfully adding the account, start updating your Google Play services:

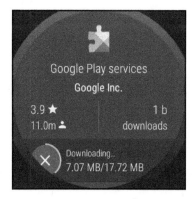

- Now, after all this configuration, compile the program in Android Studio and see the map on the Wear device:

The Google API client

`GoogleApiClient` extends the **Object** class. The Google API Client provides a common entry point to all the Google Play services and manages the network connection between the user's device and each Google service. Google recommends to use `GoogleApiClient` to get the user's location programmatically.

Create `GoogleApiClient` on each thread. `GoogleApiClient` service connections are cached internally. `GoogleApiClient` instances are not thread-safe, so creating multiple instances is fast. `GoogleApiClient` is used with a variety of static methods. Some of these methods require that `GoogleApiClient` be connected; some will queue up calls before `GoogleApiClient` is connected.

Here is a code example that creates a `GoogleApiClient` instance that connects with the Google `LocationServices`:

```
GoogleApiClient mGoogleApiClient = new GoogleApiClient.Builder(this)
.addConnectionCallbacks(this)
.addOnConnectionFailedListener(this)
.addApi(LocationServices.API)
.build();
```

Configuring the project for functionalities

We know the importance of creating packages for better code management and maintaining code in the future. Let's create a package for the project with four different names, which are adapter, model, util, and view.

We write our plain old Java objects inside the model package. We will configure all the database-related classes in the util package and the custom views, such as dialog fragments, `TextView`, and so on, in the view package. For the custom `infoWindow`, we have to create a `infoWindowAdapter` inside the `adapter` package.

It is very important to fetch the location information using `GoogleApiClient`. Now that we have configured the Wear map activity and the map is drawn with the API key that we added, it is time to work on getting location details with the help of `GoogleApiClient`.

Getting users' location information with the help of GoogleApiClient

Now, in the `MapActivity` class, we need to implement the following interfaces:

- `GoogleApiClient.ConnectionCallback`
- `GoogleApiClient.OnConnectionFailedListener`

And, we need to override three methods from these two interfaces, which are as follows:

- `public void onConnected(..){}`
- `public void onConnectionSuspended(..){}`
- `public void onConnectionFailed(..){}`

In the `onConnected` method, we can instantiate the location service using the `GoogleApiClient` instance. First, let's add the `GoogleApiClient` to the project. Create an instance of `GoogleApiClient` in the `MapActivity` global scope:

```
private GoogleApiClient mGoogleApiClient;
```

Add a void method named `addGoogleAPIClient(){ }` for retrieving the location services API:

```
private void addGoogleAPIClient(){
    mGoogleApiClient = new GoogleApiClient.Builder(this)
            .addConnectionCallbacks(this)
            .addOnConnectionFailedListener(this)
            .addApi(LocationServices.API)
            .build();
}
```

 To have the google play services to do the location related jobs, please add the following dependency in the gradle wear module:
`compile 'com.google.android.gms:play-services-location:11.0.2'`

Now, in the `onConnected` method, attach the `mGoogleApiClient`:

```
@Override
public void onConnected(@Nullable Bundle bundle) {

    Location location =     LocationServices.FusedLocationApi
    .getLastLocation(mGoogleApiClient);
    double latitude = location.getLatitude();
    double longitude = location.getLongitude();


}
```

The `Locationservice` needs a permission check before requesting the location. Let's add the permission in manifest and in Activity.

Add the following permissions to Manifest:

```
<uses-permission android:name="android.permission.INTERNET"/>
<uses-permission android:name="android.permission.ACCESS_NETWORK_STATE"/>
<uses-permission android:name="android.permission.WRITE_EXTERNAL_STORAGE"/>
<!-- The following two permissions are not required to use
    Google Maps Android API v2, but are recommended. -->
<uses-permission android:name="android.permission.ACCESS_COARSE_LOCATION"/>
<uses-permission android:name="android.permission.ACCESS_FINE_LOCATION"/>
```

Write a method in the `MapActivity.java` class for checking the permission as follows:

```
private boolean checkPermission(){
    int result = ContextCompat.checkSelfPermission(MapsActivity.this,
    Manifest.permission.ACCESS_FINE_LOCATION);
    if (result == PackageManager.PERMISSION_GRANTED){
```

```
        return true;

    } else {

        return false;

    }
}
```

Override the method `onRequestPermissionsResult(..){}` as follows:

```
@Override
public void onRequestPermissionsResult(int requestCode, String
permissions[], int[] grantResults) {
    switch (requestCode) {
        case PERMISSION_REQUEST_CODE:
            if (grantResults.length > 0 && grantResults[0] ==
            PackageManager.PERMISSION_GRANTED) {


            } else {


            }
            break;
    }
}
```

Now, we have the permission check method; handle it in the `onConnected` method:

```
@Override
public void onConnected(@Nullable Bundle bundle) {
    if (checkPermission()) {
        Location location = LocationServices.FusedLocationApi
        .getLastLocation(mGoogleApiClient);
        double latitude = location.getLatitude();
        double longitude = location.getLongitude();
    }else{
    }


}
```

Let's write a method to check whether the GPS is available `onboard` on Wear devices. Using the `packagemanager` class, we can retrieve the available hardware in a Wear device. Let's write a method called `hasGps()`:

```
private boolean hasGps() {
    return getPackageManager().hasSystemFeature(
      PackageManager.FEATURE_LOCATION_GPS);
}
```

Use this method in the `onCreate()` method if you want your users to know about whether their device has a GPS device or you just want to log it for development purposes:

```
if (!hasGps()) {
    Log.d(TAG, "This hardware doesn't have GPS.");
    // Fall back to functionality that does not use location or
    // warn the user that location function is not available.
}
```

If your wearable app records data using the built-in GPS, you may want to synchronize the location data with the handset using the `LocationListner` interface by implementing the `onLocationChanged()` method.

To make your application location-aware, use `GoogleAPIclient`.

Follow this link for understanding more about permission: `https://developer.android.com/training/articles/wear-permissions.html`.

Now, let's work with the `onMapclick` method for handling the process of adding the marker on maps. To do this, implement `GoogleMap.OnMapClickListener` in your activity and implement its callback method, which will give you the `onmapclick` with latlong. Add the click context to your `onMapReady` callback, as follows:

```
mMap.setOnMapClickListener(this);
```

In the `onMapClick` method, we can add the following marker using `latLng`:

```
@Override
public void onMapClick(LatLng latLng) {
    Log.d(TAG, "Latlng is "+latLng);
}
```

Adding a marker in the `onMapclick` method uses `MarkerOptions()`. For custom markers designed by Google, we will use the `addmarker` method from maps and add the new `MarkerOptions` with position, title, and snippet, which is minimal description below the title:

```
@Override
public void onMapClick(LatLng latLng) {
    Log.d(TAG, "Latlng is "+latLng);
    mMap.addMarker(new MarkerOptions()
            .position(latLng)
            .title("Packt wear 2.0")
            .snippet("Map is cool in wear device"));
}
```

After adding the marker with `infowindow`:

Now, we have the map and we are adding the marker to the map, but we need to work with geocoding for fetching the address name of the coordinates.

GeoSpatial data using GeoCoder

Fetch the address using coordinates using the `GeoCoder` class. Geocoding, usually, is a process of transforming a street address or other description of a location into a (latitude, longitude) coordinate. Reverse geocoding is the process of transforming a (latitude, longitude) coordinate into a (partial) address.

In the `OnMapClick` method, make the following changes:

```
@Override
public void onMapClick(LatLng latLng) {
    Log.d(TAG, "Latlng is "+latLng);
    //Fetching the best address match
    Geocoder geocoder = new Geocoder(this);
    List<Address> matches = null;
    try {
        matches = geocoder.getFromLocation(latLng.latitude,
        latLng.longitude, 1);
    } catch (IOException e) {
        e.printStackTrace();
    }
    Address bestAddress = (matches.isEmpty()) ? null : matches.get(0);
    int maxLine = bestAddress.getMaxAddressLineIndex();

    mMap.addMarker(new MarkerOptions()
            .position(latLng)
            .title(bestAddress.getAddressLine(maxLine - 1))
            .snippet(bestAddress.getAddressLine(maxLine)));
}
```

The preceding code snippet adds the marker to the map with the location name in the info window:

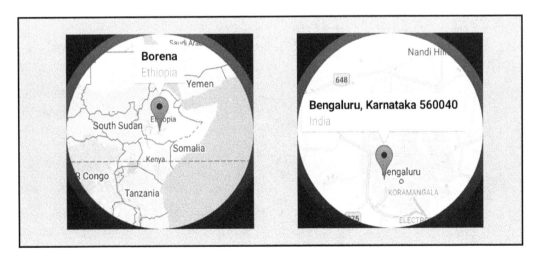

The pop-up view on clicking the map is called `infowindow` in Android. It is a similar component to ToolTip in web development. In this project, we need to save the data wherever users click on the map; we need to show a custom map marker with the help of `infowindow`. We need to write an adapter implementing `GoogleMap.InfoWindowAdapter` with a custom layout, as follows:

Infowindow adapter

The following implementation explains how to write our custom `infowindow` adapter for the map marker:

```xml
//XML latout for customising  infowindow
<LinearLayout xmlns:android="http://schemas.android.com/apk/res/android"
              android:orientation="horizontal"
              android:layout_width="match_parent"
              android:layout_height="match_parent">
    <TextView
        android:id="@+id/title"
        android:layout_width="wrap_content"
        android:layout_height="wrap_content"/>
    <TextView
        android:id="@+id/snippet"
        android:layout_width="wrap_content"
        android:layout_height="wrap_content"/>
</LinearLayout>
```

The `windowadapter` class implements `GoogleMap.InfoWindowAdapter` with two callback methods `getInfoWindow(..){}` and `getInfoContents(..){}`. We can inflate our custom layout `getInfoContent` method:

```java
public class WearInfoWindowAdapter implements GoogleMap.InfoWindowAdapter {
    private LayoutInflater mLayoutInflater;
    private View mView;
    MarkerAdapter(LayoutInflater layoutInflater){
        mLayoutInflater = layoutInflater;
    }
    @Override
    public View getInfoWindow(Marker marker) {
        return null;
    }
    @Override
    public View getInfoContents(Marker marker) {
        if (mView == null){
            mView = mLayoutInflater.inflate(R.layout.marker, null);
        }
```

```
    TextView titleView = (TextView)mView.findViewById(R.id.title);
    titleView.setText(marker.getTitle());
    TextView snippetView =
    (TextView)mView.findViewById(R.id.snippet);
    snippetView.setText(marker.getSnippet());
    return mView;
    }
}
```

Add the preceding adapter class to the adapter package for better code access and maintenance. `InfoWindowAdapter` does not use any data to populate the view; we populate the view with whatever data is associated with the marker. If we want to add anything beyond the title and the snippet, we have no way of doing it in the adapter itself. We need to create a mechanism for achieving this programmatically.

Create the `Memory` class in the model package. The `Memory` class is a place where users choose to add the marker:

```
public class Memory {
    double latitude;
    double longitude;
    String city; // City name
    String country; // Country name
    String notes; // saving notes on the location
}
```

Now, we have our memory and the custom `infowindow` adapter is ready for working with the `onMapclick` implementation. For each marker, we shall add a memory class association. To save all our memory temporarily, let's use `HashMap`:

```
private HashMap<String, Memory> mMemories = new HashMap<>();
```

Let's add our marker to `HashMap` for access to `Marker` properties, such as `Marker` ID and so on. The complete code for the adapter is as follows:

```
public class WearInfoWindowAdapter implements GoogleMap.InfoWindowAdapter {

    public LayoutInflater mLayoutInflater;
    public View mView;
    public HashMap<String, Memory> mMemories;

    WearInfoWindowAdapter(LayoutInflater layoutInflater,
    HashMap<String,Memory> memories){
        mLayoutInflater = layoutInflater;
        mMemories = memories;
    }
```

```
@Override
public View getInfoWindow(Marker marker) {
    return null;
}

@Override
public View getInfoContents(Marker marker) {
    if (mView == null) {
        mView = mLayoutInflater.inflate(R.layout.marker, null);
    }
    Memory memory = mMemories.get(marker.getId());

    TextView titleView = (TextView)mView.findViewById(R.id.title);
    titleView.setText(memory.city);
    TextView snippetView =
    (TextView)mView.findViewById(R.id.snippet);
    snippetView.setText(memory.country);
    TextView notesView = (TextView)mView.findViewById(R.id.notes);
    notesView.setText(memory.notes);

    return mView;
}
}
```

In the OnMapClick method, add the following changes:

```
@Override
public void onMapClick(LatLng latLng) {
    Log.d(TAG, "Latlng is "+latLng);

    Geocoder geocoder = new Geocoder(this);
    List<Address> matches = null;
    try {
        matches = geocoder.getFromLocation(latLng.latitude,
        latLng.longitude, 1);
    } catch (IOException e) {
        e.printStackTrace();
    }

    Address bestAddress = (matches.isEmpty()) ? null : matches.get(0);
    int maxLine = bestAddress.getMaxAddressLineIndex();

    Memory memory = new Memory();
    memory.city = bestAddress.getAddressLine(maxLine - 1);
    memory.country = bestAddress.getAddressLine(maxLine);
    memory.latitude = latLng.latitude;
    memory.longitude = latLng.longitude;
    memory.notes = "Packt and wear 2.0 notes...";
```

```
    Marker marker = mMap.addMarker(new MarkerOptions()
            .position(latLng));

    mMemories.put(marker.getId(), memory);
}
```

Attach the new `Marker` to the maps with the following code in the `onMapready` method:

```
mMap.setInfoWindowAdapter(new WearInfoWindowAdapter(getLayoutInflater(),
mMemories));
```

Now, compile the program. You should be able to see the updated `infoWindow`, as follows:

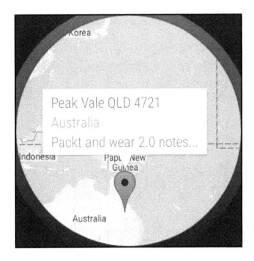

Custom DialogFragment for taking notes about the location

`DialogFragment` is a dialog window that floats in activity. On Wear devices it will not float but it gives the Wear optimized design. Check the following code for implementing.

Before moving forward, implement the Memory class to a serializable interface:

```
public class Memory implements Serializable {
    public double latitude;
    public double longitude;
    public String city;
    public String country;
    public String notes;
}
```

Add the following layout file to the layout directory and call the layout file as `memory_dialog_fragment.xml`. After creating the file, add the following code inside the layout file:

```xml
<?xml version="1.0" encoding="utf-8"?>
<android.support.wearable.view.BoxInsetLayout
    xmlns:android="http://schemas.android.com/apk/res/android"
    xmlns:tools="http://schemas.android.com/tools"
    xmlns:app="http://schemas.android.com/apk/res-auto"
    android:id="@+id/container"
    android:layout_width="match_parent"
    android:layout_height="match_parent"
    tools:context=".MapsActivity"
    tools:deviceIds="wear">

    <LinearLayout
        xmlns:android="http://schemas.android.com/apk/res/android"
        android:layout_width="match_parent"
        android:layout_height="wrap_content"
        android:orientation="vertical"
        android:padding="5dp"
        app:layout_box="all"
        android:layout_gravity="center"
        android:gravity="center">


        <TextView
            android:id="@+id/city"
            android:layout_width="wrap_content"
            android:layout_height="wrap_content" />

        <TextView
            android:id="@+id/country"
            android:layout_width="wrap_content"
            android:layout_height="wrap_content" />

        <EditText
            android:id="@+id/notes"
            android:layout_width="match_parent"
            android:layout_height="wrap_content" />

    </LinearLayout>
</android.support.wearable.view.BoxInsetLayout>
```

After creating the layout file, let's work on the Java code for creating the custom dialog. Create a class called `MemoryDialogFragment` and extend it to `DialogFragment`.

Create an interface for handling the `SaveClicked` and `cancelClicked` buttons of `DialogFragment`:

```
public interface Listener{
    public void OnSaveClicked(Memory memory);
    public void OnCancelClicked(Memory memory);
}
```

Now, add the following instances to the `MemoryDialogFragment` global scope.:

```
private static final String TAG = "MemoryDialogFragment";
private static final String MEMORY_KEY = "MEMORY";
private Memory mMemory;
private Listener mListener;
private View mView;
```

Now, work on inflating the layout with the right data in the right field:

```
@Override
public Dialog onCreateDialog(Bundle savedInstanceState) {

    mView = getActivity().getLayoutInflater()
    .inflate(R.layout.memory_dialog_fragment, null);
    TextView cityView = (TextView) mView.findViewById(R.id.city);
    cityView.setText(mMemory.city);
    TextView countryView = (TextView) mView.findViewById(R.id.country);
    countryView.setText(mMemory.country);

    AlertDialog.Builder builder = new
    AlertDialog.Builder(getActivity());
    builder.setView(mView)
            .setTitle(getString(R.string.dialog_title))
            .setPositiveButton(getString(R.string.DialogSaveButton),
            new DialogInterface.OnClickListener() {
                @Override
                public void onClick(DialogInterface dialog, int which)
                {
                    EditText notesView = (EditText)
                    mView.findViewById(R.id.notes);
                    mMemory.notes = notesView.getText().toString();
                    mListener.OnSaveClicked(mMemory);
                }
            })
            .setNegativeButton(getString(R.string.DialogCancelButton),
            new DialogInterface.OnClickListener() {
                @Override
                public void onClick(DialogInterface dialog, int which)
                {
```

```
                    mListener.OnCancelClicked(mMemory);
                }
            });

        return builder.create();
    }
```

We will fetch the serialized data from `Memory` in the `oncreate` method:

```
@Override
public void onCreate(Bundle savedInstanceState) {
    super.onCreate(savedInstanceState);
    Bundle args = getArguments();
    if (args != null){
        mMemory = (Memory)args.getSerializable(MEMORY_KEY);
    }
}
```

The complete code for the `MemoryDialogFragment` is as follows:

```
public class MemoryDialogFragment extends DialogFragment   {

    private static final String TAG = "MemoryDialogFragment";
    private static final String MEMORY_KEY = "MEMORY";

    private Memory mMemory;
    private Listener mListener;
    private View mView;


    @Override
    public void onCreate(Bundle savedInstanceState) {
        super.onCreate(savedInstanceState);
        Bundle args = getArguments();
        if (args != null){
            mMemory = (Memory)args.getSerializable(MEMORY_KEY);
        }
    }

    @Override
    public Dialog onCreateDialog(Bundle savedInstanceState) {

        mView = getActivity().getLayoutInflater()
        .inflate(R.layout.memory_dialog_fragment, null);
        TextView cityView = (TextView) mView.findViewById(R.id.city);
        cityView.setText(mMemory.city);
        TextView countryView = (TextView)
        mView.findViewById(R.id.country);
```

```
        countryView.setText(mMemory.country);

        AlertDialog.Builder builder = new
        AlertDialog.Builder(getActivity());
        builder.setView(mView)
                .setTitle(getString(R.string.dialog_title))
                .setPositiveButton(getString
                (R.string.DialogSaveButton),
                new DialogInterface.OnClickListener() {
                    @Override
                    public void onClick
                    (DialogInterface dialog, int which) {
                        EditText notesView = (EditText)
                        mView.findViewById(R.id.notes);
                        mMemory.notes = notesView.getText().toString();
                        mListener.OnSaveClicked(mMemory);
                    }
                })
                .setNegativeButton(getString
                (R.string.DialogCancelButton),
                new DialogInterface.OnClickListener() {
                    @Override
                    public void onClick
                    (DialogInterface dialog, int which) {
                        mListener.OnCancelClicked(mMemory);
                    }
                });

        return builder.create();
}

public static MemoryDialogFragment newInstance(Memory memory){
    MemoryDialogFragment fragment = new MemoryDialogFragment();
    Bundle args = new Bundle();
    args.putSerializable(MEMORY_KEY, memory);
    fragment.setArguments(args);

    return fragment;
}

@Override
public void onAttach(Activity activity) {
    super.onAttach(activity);
    try{
        mListener = (Listener)getActivity();
    }catch (ClassCastException e){
        throw new IllegalStateException("Activity does not
        implement contract");
```

```
            }

        }

        @Override
        public void onDetach() {
            super.onDetach();
            mListener = null;
        }

        public interface Listener{
            public void OnSaveClicked(Memory memory);
            public void OnCancelClicked(Memory memory);
        }
    }
```

In the `OnMapClick` method, make the following changes:

```
@Override
public void onMapClick(LatLng latLng) {
    Log.d(TAG, "Latlng is "+latLng);

    Memory memory = new Memory();
    updateMemoryPosition(memory, latLng);
    MemoryDialogFragment.newInstance(memory)
    .show(getFragmentManager(),MEMORY_DIALOG_TAG);
}
```

Now, compile the program. You will see the following screen on `mapclick`. The user can input his or her thoughts about the map location in the edittext field:

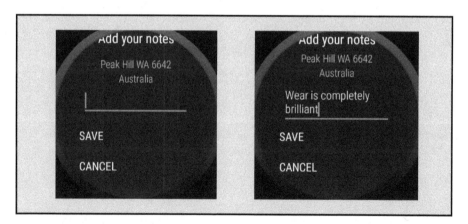

Now that we have added the dialog for taking the input, let's work on saving the data to SQLite.

Configuring SQLite and saving the markers

Persisting all the necessary data is the fundamental use case for any good software. Android SDK provides an SQLite storage solution built in. It has a very small footprint and is very fast. If a programmer is familiar with SQL queries and operations, SQLite is going to be easy and delightful to work with.

Schema and contract

Essentially, for a database, we need to create a data schema, which is a formal declaration of how the database is organized. The schema is reflected in the SQLite query statements. A contract class is a container for constants that define names for URIs, tables, and columns. The contract class allows the use of the same constants across all the other classes in the same package.

For the scope of `WearMapDiary`, we will create all the instances in the `DBHelper` class. Now, let's create the `DBhelper` class, which opens and connects the application to SQLite and processes the query:

```
public class DbHelper extends SQLiteOpenHelper {

    private static final String DATABASE_NAME = "traveltracker.db";
    private static final int DATABASE_VERSION = 3;
    public static final String MEMORIES_TABLE = "memories";
    public static final String COLUMN_LATITUDE = "latitude";
    public static final String COLUMN_LONGITUDE = "longitude";
    public static final String COLUMN_CITY = "city";
    public static final String COLUMN_COUNTRY = "country";
    public static final String COLUMN_NOTES = "notes";
    public static final String COLUMN_ID = "_id";

    private static DbHelper singleton = null;

    public static DbHelper getInstance(Context context){
        if (singleton == null){
            singleton = new DbHelper(context.getApplicationContext());
        }
        return singleton;
    }
```

```
        private DbHelper(Context context) {
            super(context, DATABASE_NAME, null, DATABASE_VERSION);
        }

        @Override
        public void onCreate(SQLiteDatabase db) {
            db.execSQL("CREATE TABLE "+MEMORIES_TABLE+" ("
                    +COLUMN_ID+" INTEGER PRIMARY KEY AUTOINCREMENT, "
                    +COLUMN_LATITUDE +" DOUBLE, "
                    +COLUMN_LONGITUDE +" DOUBLE, "
                    +COLUMN_CITY +" TEXT, "
                    +COLUMN_COUNTRY +" TEXT, "
                    +COLUMN_NOTES +" TEXT"
                    +")");
        }

        @Override
        public void onUpgrade(SQLiteDatabase db, int oldVersion, int
        newVersion) {
            db.execSQL("DROP TABLE IF EXISTS "+MEMORIES_TABLE);
            onCreate(db);
        }
    }
```

We need to create a `Datasource` for managing all the queries, and reading and writing the data to SQLite. Here, in this class, we will create multiple methods for creating data, reading the data, updating the data, and deleting the data:

```
public class MemoriesDataSource {
    private DbHelper mDbHelper;
    private String[] allColumns = {
            DbHelper.COLUMN_ID, DbHelper.COLUMN_CITY,
            DbHelper.COLUMN_COUNTRY, DbHelper.COLUMN_LATITUDE,
            DbHelper.COLUMN_LONGITUDE, DbHelper.COLUMN_NOTES
    };

    public MemoriesDataSource(Context context){
        mDbHelper = DbHelper.getInstance(context);
    }

    public void createMemory(Memory memory){
        ContentValues values = new ContentValues();
        values.put(DbHelper.COLUMN_NOTES, memory.notes);
        values.put(DbHelper.COLUMN_CITY, memory.city);
        values.put(DbHelper.COLUMN_COUNTRY, memory.country);
        values.put(DbHelper.COLUMN_LATITUDE, memory.latitude);
        values.put(DbHelper.COLUMN_LONGITUDE, memory.longitude);
        memory.id = mDbHelper.getWritableDatabase()
```

```
        .insert(DbHelper.MEMORIES_TABLE, null, values);
}

public List<Memory> getAllMemories(){

    Cursor cursor = allMemoriesCursor();
    return cursorToMemories(cursor);
}

public Cursor allMemoriesCursor(){
    return mDbHelper.getReadableDatabase()
    .query(DbHelper.MEMORIES_TABLE,
    allColumns,null, null, null, null, null);
}

public List<Memory> cursorToMemories(Cursor cursor){
    List<Memory> memories =  new ArrayList<>();
    cursor.moveToFirst();
    while (!cursor.isAfterLast()){
        Memory memory = cursorToMemory(cursor);
        memories.add(memory);
        cursor.moveToNext();
    }
    return memories;
}

public void updateMemory(Memory memory){
    ContentValues values = new ContentValues();
    values.put(DbHelper.COLUMN_NOTES, memory.notes);
    values.put(DbHelper.COLUMN_CITY, memory.city);
    values.put(DbHelper.COLUMN_COUNTRY, memory.country);
    values.put(DbHelper.COLUMN_LATITUDE, memory.latitude);
    values.put(DbHelper.COLUMN_LONGITUDE, memory.longitude);

    String [] whereArgs = {String.valueOf(memory.id)};

    mDbHelper.getWritableDatabase().update(
            mDbHelper.MEMORIES_TABLE,
            values,
            mDbHelper.COLUMN_ID+"=?",
            whereArgs
    );
}

public void deleteMemory(Memory memory){
    String [] whereArgs = {String.valueOf(memory.id)};

    mDbHelper.getWritableDatabase().delete(
```

```
                        mDbHelper.MEMORIES_TABLE,
                        mDbHelper.COLUMN_ID+"=?",
                        whereArgs
            );
        }

    private Memory cursorToMemory(Cursor cursor){
        Memory memory = new Memory();
        memory.id = cursor.getLong(0);
        memory.city = cursor.getString(1);
        memory.country = cursor.getString(2);
        memory.latitude = cursor.getDouble(3);
        memory.longitude = cursor.getDouble(4);
        memory.notes = cursor.getString(5);
        return memory;
    }
}
```

For executing all these queries in the background using `cursorLoader`, we will write another class and we will call this class `DBCurserLoader`:

```
public abstract class DbCursorLoader extends AsyncTaskLoader<Cursor> {

    private Cursor mCursor;

    public DbCursorLoader(Context context){
        super(context);
    }

    protected abstract Cursor loadCursor();

    @Override
    public Cursor loadInBackground() {
        Cursor cursor = loadCursor();
        if (cursor != null){
            cursor.getCount();
        }

        return cursor;
    }

    @Override
    public void deliverResult(Cursor data) {
        Cursor oldCursor = mCursor;
        mCursor = data;

        if (isStarted()){
            super.deliverResult(data);
```

```
    }

    if (oldCursor != null && oldCursor != data){
        onReleaseResources(oldCursor);
    }
}

@Override
protected void onStartLoading() {
    if (mCursor != null){
        deliverResult(mCursor);
    }
    if (takeContentChanged() || mCursor == null){
        forceLoad();
    }
}

@Override
protected void onStopLoading() {
    cancelLoad();
}

@Override
public void onCanceled(Cursor data) {
    super.onCanceled(data);

    if (data != null) {
        onReleaseResources(data);
    }
}

@Override
protected void onReset() {
    super.onReset();

    onStopLoading();

    if (mCursor != null){
        onReleaseResources(mCursor);
    }
    mCursor = null;
}

private void onReleaseResources(Cursor cursor){
    if (!cursor.isClosed()){
        cursor.close();
    }
}
```

```
    }
```

Create another class for loading all the memories from `memoryDatasource` and extend it to `DBCursorLoader`:

```
public class MemoriesLoader extends DbCursorLoader {

    private MemoriesDataSource mDataSource;

    public MemoriesLoader(Context context, MemoriesDataSource
    memoriesDataSource){
        super(context);
        mDataSource = memoriesDataSource;
    }

    @Override
    protected Cursor loadCursor() {
        return mDataSource.allMemoriesCursor();
    }
}
```

Now, we have the SQLite configuration working fine. Lets work with `MapActivity` to save the data inside SQLite `onMapclick`.

Saving data in SQLite

To connect SQLite to the activity and to save the data in SQLite, implement the activity `LoaderManager.LoaderCallbacks<Cursor>` and instatiate the datasource in the `onCreate` method:

```
mDataSource = new MemoriesDataSource(this);
getLoaderManager().initLoader(0,null,this);
```

Implement the callback methods for the `LoaderManager.LoaderCallbacks<Cursor>` interface:

```
@Override
public Loader<Cursor> onCreateLoader(int id, Bundle args) {
    return null;
}

@Override
public void onLoadFinished(Loader<Cursor> loader, Cursor data) {

}
```

```
@Override
public void onLoaderReset(Loader<Cursor> loader) {

}
```

Now, refactor the `addingMarker` code in a method as follows:

```
private void addMarker(Memory memory) {
    Marker marker = mMap.addMarker(new MarkerOptions()
            .draggable(true)
            .position(new LatLng(memory.latitude, memory.longitude)));

    mMemories.put(marker.getId(), memory);
}
```

We still need to work with dragging the marker for future implementation. Let's make the draggable property true. Now, in the `OnMapClick` method, call the following code:

```
@Override
public void onMapClick(LatLng latLng) {
    Log.d(TAG, "Latlng is "+latLng);

    Memory memory = new Memory();
    updateMemoryPosition(memory, latLng);
    MemoryDialogFragment.newInstance(memory)
    .show(getFragmentManager(),MEMORY_DIALOG_TAG);
}
```

Let's refactor the `UpdateMemoryPosition` method, which fetches the address from `latlng` and adds it to `Memory`:

```
private void updateMemoryPosition(Memory memory, LatLng latLng) {
    Geocoder geocoder = new Geocoder(this);
    List<Address> matches = null;
    try {
        matches = geocoder.getFromLocation(latLng.latitude,
        latLng.longitude, 1);
    } catch (IOException e) {
        e.printStackTrace();
    }

    Address bestMatch = (matches.isEmpty()) ? null : matches.get(0);
    int maxLine = bestMatch.getMaxAddressLineIndex();
    memory.city = bestMatch.getAddressLine(maxLine - 1);
    memory.country = bestMatch.getAddressLine(maxLine);
    memory.latitude = latLng.latitude;
    memory.longitude = latLng.longitude;
}
```

Now, we are saving the data inside SQLite. When we close and open the map, we are not reading and adding the marker data to the map:

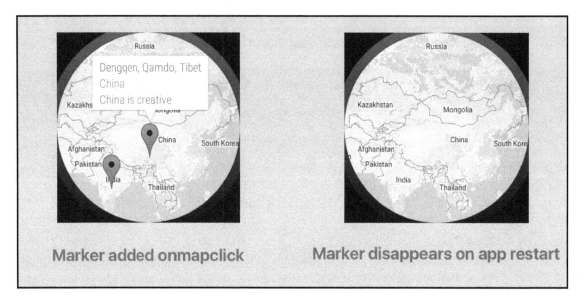

Now, let's read the SQLite data and add it to the maps.

The `onCreateLoader` callback method from the `LoaderManager` class adds the data through the `Datasource` instance to `MemoryLoader` as follows:

```
@Override
public Loader<Cursor> onCreateLoader(int id, Bundle args) {
    Log.d(TAG,"onCreateLoader");
    return new MemoriesLoader(this, mDataSource);
}
```

In the `onLoadFinished` method, fetch the data from the cursor and add it to the maps:

```
@Override
public void onLoadFinished(Loader<Cursor> loader, Cursor cursor) {
    Log.d(TAG,"onLoadFinished");
    onFetchedMemories(mDataSource.cursorToMemories(cursor));
}
```

Adding the marker to map from fetched data:

```
private void onFetchedMemories(List<Memory> memories) {
    for(Memory memory: memories){
        addMarker(memory);
    }
}
```

Now, we have a functional Wear app called `WearMapDiary`, which finds the address and saves quick notes about the location in the map. It adds the marker to the SQLite database and attaches the marker to the map when we open the application in a Wear device:

Retrieved notes and location information from the app:

In this dialog, users can type in the data he or she wants to save in the current location:

So far, we have explored how to integrate maps to Wear devices and had a clear understanding of getting the Maps API key. We are using `GoogleApiclient` for fetching the location service. We are checking for GPS hardware availability:

Following steps briefly, explains how to write the custom marker.

- Explored the custom `InfoWindow` adapter by implementing `GoogleMap.InfoWindowAdapter`
- Created a custom `dialogFragment` for Wear compatibility using `boxinsetlayout`
- `Geocoder` class fetching the `GeoSpatial` data
- SQLite and its integration for map data

Now, it's time to understand more about maps on Wear devices.

Difference between a standalone map and mobile-synced map application

A watch application that is target at the Wear 2.0 platform can connect to Wi-Fi through the onboard Wi-Fi transceiver. We can cache the maps and a lot more, but it still lacks the comforts that a mobile map application gives. Generally, for standalone Wear applications, the target API level is 25 and comes with runtime permissions for security operations. In this chapter, we have added the code for handling runtime permissions.

Identifying the app as standalone

Wear 2.0 requires a new metadata element in the Android `Manifest` file of watch apps, as a child of the `<application>` element. The name of the new metadata element is `com.google.android.wearable.standalone` and the value must be true or false:

```
<meta-data
    android:name="com.google.android.wearable.standalone"
    android:value="true" />
```

Since standalone apps are independent or semi-independent, it can be installed by an iPhone user and Android phone that lacks the Play store (BlackBerry android forked OS and Nokia custom Android phones).

If a watch app depends on a phone app, set the value of the previous metadata element to false.

Even if the value is false, the watch app can be installed before the corresponding phone app is installed. Therefore, if a watch app detects that a companion phone lacks a necessary phone app, the watch app should prompt the user to install the phone app.

Sharing data between a Wear app and phone app

Data can be shared between a Wear application and a phone application or data that is specific to an application. You can use standard Android storage APIs to store data locally. For example, you can use `SharedPreferences APIs`, SQLite, or internal storage (as you would in the case of a phone). The Messaging API Wear app can communicate with the corresponding phone application.

Detecting your application from another device

In **CapabilityAPI**, your Wear application can detect the corresponding mobile application for a Wear application. Wear devices can advertise their events to the paired device spontaneously and statically as well. For checking the advertised capabilities of paired Wear devices, check this link for more information: `https://developer.android.com/tra ining/wearables/data-layer/messages.html#AdvertiseCapabilities`.

Note that not all the phones support the Play store (such as iPhones, and so on). This section describes the best practices for these scenarios: your standalone watch app needs your phone app and your phone app needs your standalone watch app.

Specifying capability names to detect your apps

For the app corresponding to each device type (watch or phone), specify a unique string for the capability name in the `res/values/wear.xml` file. For example, in your mobile module, the `wear.xml` file could include the following code in Wear and mobile modules:

```
<resources>
    <string-array name="android_wear_capabilities">
        <item>verify_remote_example_phone_app</item>
    </string-array>
</resources>
```

Detecting and guiding the user to install a corresponding phone app

Wear 2.0 introduces standalone applications. Wear apps are capable enough to function without the support of a mobile app. In a critical situation, when a mobile app is a must, a Wear app can guide the user to install the mobile support app and the Wear correspondent app with the following set of steps:

- Use `CapabilityApi` to check whether your phone app is installed on the paired phone. For more information, check this sample by Google: `https://github.com/googlesamples/android-WearVerifyRemoteApp`.

- If your phone app isn't installed on the phone, use `PlayStoreAvailability.getPlayStoreAvailabilityOnPhone()` to check what type of phone it is.

- If `PlayStoreAvailability.PLAY_STORE_ON_PHONE_AVAILABLE` is returned `true` which addresses that Playstore is installed in the phone.

- Call `RemoteIntent.startRemoteActivity()` on the Wear device to open the Play Store on the phone using the market URI (`market://details?id=com.example.android.wearable.wear.finddevices`).

- If `PlayStoreAvailability.PLAY_STORE_ON_PHONE_UNAVAILABLE` is returned, it means that the phone is likely an iOS phone (with no Play Store available). Open the App Store on the iPhone by calling `RemoteIntent.startRemoteActivity()` on the Wear device with this URI: `https://itunes.apple.com/us/app/yourappname`. Also, see opening a URL from a watch. On an iPhone, from Android Wear, you cannot programmatically determine if your phone app is installed. As a best practice, provide a mechanism to the user (for example, a button) to manually trigger the opening of the App Store.

For a more detailed understanding about standalone applications, do check out the following link: `https://developer.android.com/wear/preview/features/standalone-apps.html`.

Keeping your application active on a Wear device

When we write an application for different contexts, we need to do certain alternatives. We know that when not using the application, we should make that app to sleep in Wear devices for better battery performance, but when we are building an application for maps, it is necessary that maps be visible and active for the user.

Android has a simple configuration for this: a method with few lines that activates the ambient mode:

```
//oncreate Method
setAmbientEnabled();
```

This starts the ambient mode on the map. The API swaps to a non-interactive and low-color rendering of the map when the user is no longer actively using the app:

```
@Override
public void onEnterAmbient(Bundle ambientDetails) {
    super.onEnterAmbient(ambientDetails);
    mMapFragment.onEnterAmbient(ambientDetails);
}
```

The following code exits the ambient mode on the WearMap. The API swaps to the normal rendering of the map when the user starts actively using the app:

```
@Override
public void onEnterAmbient(Bundle ambientDetails) {
    super.onEnterAmbient(ambientDetails);
    mMapFragment.onEnterAmbient(ambientDetails);
}
```

Configuring WAKE_LOCK for your application

Few Wear applications are very useful when they are constantly visible. Making an app constantly visible has an impact on battery life, so you should carefully consider that impact when adding this feature to your app.

Add the WAKE_LOCK permission to manifest:

```
<uses-permission android:name="android.permission.WAKE_LOCK" />
```

The following code snippet helps to understand the WAKE_LOCK mechanism:

```
// Schedule a new alarm
    if (isAmbient()) {
        // Calculate the next trigger time
        long delayMs = AMBIENT_INTERVAL_MS - (timeMs %
        AMBIENT_INTERVAL_MS);
        long triggerTimeMs = timeMs + delayMs;

        mAmbientStateAlarmManager.setExact(
            AlarmManager.RTC_WAKEUP,
            triggerTimeMs,
            mAmbientStatePendingIntent);

    } else {
        // Calculate the next trigger time for interactive mode
    }
```

Instead of using the Input method framework for reading the input, users can also use the voice input, which requires active internet on your Wear device:

```
private static final int SPEECH_REQUEST_CODE = 0;

// Create an intent that can start the Speech Recognizer activity
private void displaySpeechRecognizer() {
    Intent intent = new
    Intent(RecognizerIntent.ACTION_RECOGNIZE_SPEECH);
```

```
    intent.putExtra(RecognizerIntent.EXTRA_LANGUAGE_MODEL,
            RecognizerIntent.LANGUAGE_MODEL_FREE_FORM);
// Start the activity, the intent will be populated with the speech text
    startActivityForResult(intent, SPEECH_REQUEST_CODE);
}

// This callback is invoked when the Speech Recognizer returns.
// This is where you process the intent and extract the speech text from
the intent.
@Override
protected void onActivityResult(int requestCode, int resultCode,
        Intent data) {
    if (requestCode == SPEECH_REQUEST_CODE && resultCode == RESULT_OK)
    {
        List<String> results = data.getStringArrayListExtra(
                RecognizerIntent.EXTRA_RESULTS);
        String spokenText = results.get(0);
        // Do something with spokenText
    }
    super.onActivityResult(requestCode, resultCode, data);
}
```

Understanding fully interactive mode and lite mode

The Google Maps android API can serve static images as light mode maps.

Adding Lite mode to Android Maps is similar to configuring the normal maps, because it will use the same classes and interfaces. We can set Google Maps to the Lite mode in the following two ways:

- As an XML attribute to your `MapView` or `MapFrgament`
- Using the `GoogleMapOptions` object

```
<fragment xmlns:android="http://schemas.android.com/apk/res/android"
    xmlns:map="http://schemas.android.com/apk/res-auto"
    android:name="com.google.android.gms.maps.MapFragment"
    android:id="@+id/map"
    android:layout_width="match_parent"
    android:layout_height="match_parent"
    map:cameraZoom="13"
    map:mapType="normal"
    map:liteMode="true"/>
```

Or, using the `GoogleMapOptions` object as follows:

```
GoogleMapOptions options = new GoogleMapOptions().liteMode(true);
```

Interactive mode allows the application to use all the lifecycle methods, including `onCreate()`, `onDestroy()`, `onResume()`, and `onPause()`, and all the Google API features to make the application fully interactive. As a cost, there will be a network dependency.

For more information about Interactive and Lite mode, check the following link: `https://developers.google.com/maps/documentation/android-api/lite`.

Summary

Here, we are at the chapter's end, looking forward to what improvements we can do in the WearMapDiary app. Now, we know about creating a `MapsActivity`, setting up the maps and Google API key, configuring Google Play services in a Wear emulator, runtime permissions check, checking for the GPS hardware, and fetching the location name using the `geocoder` class. We have understood the concept of the interactive mode and Lite mode for maps. In the next chapter, let's understand more Wear and map UI controls and other Google Map technologies, such as streetview, changing the map types, and so on.

7

Ways to Get Around Anywhere
- UI controls and More

Now that you have learned how to bring life to Google Maps for Android Wear applications and explore SQLite integration, we need UI controls and more enhancement. In this chapter, let's focus on making the map application more functional and intuitive by adding features, such as moving the marker on the map and changing the map types. You will learn the following topics in this chapter:

- Marker controls
- Map types
- Map zoom controls
- Streetview on Wear devices
- Best practices

The marker is not only a symbol that denotes the coordinates on the map. Markers are utilized to convey what sort of place it is by replacing the marker default symbol with significant pictorial portrayals; for instance, if it's a fuel station, the marker symbol can be a little like a fuel gun symbol or hospital.

Changing marker color and customizing

Using the `MarkerOptions` class, we can change the color and icon of the marker. The following code explains both changing the icon and color of the marker.

To change the color of the marker, please refer to the following code:

```
private void addMarker(Memory memory) {
    Marker marker = mMap.addMarker(new MarkerOptions()
            .draggable(true).icon(BitmapDescriptorFactory.defaultMarker
            (BitmapDescriptorFactory.HUE_CYAN)).alpha(0.7f)
            .position(new LatLng(memory.latitude, memory.longitude)));

    mMemories.put(marker.getId(), memory);
}
```

We can now see that the marker color is changed from red to cyan with transparency. If you wish to remove the transparency, you can remove the `.alpha()` value passed to marker options:

Changed marker color

To change the marker to an icon inside the `drawable` directory, check the following code:

```
private void addMarker(Memory memory) {
    Marker marker = mMap.addMarker(new MarkerOptions()
            .draggable(true).icon(BitmapDescriptorFactory.fromResource
            (R.drawable.ic_edit_location)).alpha(0.7f)
            .position(new LatLng(memory.latitude, memory.longitude)));

    mMemories.put(marker.getId(), memory);
}
```

This replaces the default marker icon with the custom image that we are passing from a drawable directory. We need to ensure the icon size is not bloated and it has the optimal size of 72x72:

Changed marker image with custom image

The previous code snippet will help in changing the color or icon of the marker, but for more complex scenarios, we can dynamically build the marker visual asset and add it to the maps.

How about having our own custom designed marker using simple Java code? We shall create a marker with simple text drawn on top of the image. The following code explains how can we use the `Bitmap` class and `Canvas` class to draw text on top of the image:

```
private void addMarker(Memory memory) {

    Bitmap.Config conf = Bitmap.Config.ARGB_8888;
    Bitmap bmp = Bitmap.createBitmap(80, 80, conf);
    Canvas canvas1 = new Canvas(bmp);

    // paint defines the text color, stroke width and size
    Paint color = new Paint();
    color.setTextSize(15);
    color.setColor(Color.BLACK);

    // modify canvas
    canvas1.drawBitmap(BitmapFactory.decodeResource(getResources(),
            R.drawable.ic_edit_location), 0,0, color);
    canvas1.drawText("Notes", 30, 35, color);
```

```
                // add marker to Map


        Marker marker = mMap.addMarker(new MarkerOptions()
                .draggable(true).icon(BitmapDescriptorFactory
                .fromBitmap(bmp)).alpha(0.7f)
                .position(new LatLng(memory.latitude,
                memory.longitude)));

        mMemories.put(marker.getId(), memory);
    }
```

The following screenshot shows the marker with Notes drawn using bitmap and canvas:

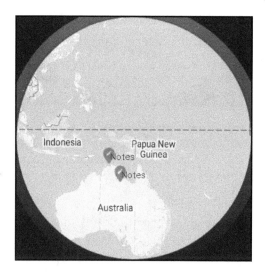

Dynamically adding the marker from realmdb information.

Dragging the marker and updating the location

Implement the GoogleMap.OnMarkerDragListener interfaces in MapActivity and implement all the callback methods from the OnMarkerDragListener interface:

```
@Override
public void onMarkerDragStart(Marker marker) {
}

@Override
public void onMarkerDrag(Marker marker) {
```

```
}

@Override
public void onMarkerDragEnd(Marker marker) {
}
```

After implementing these three methods from the interface, in the third callback `onMarkerDragEnd`, we can update the memory with the updated location details. We can also register `draglistner` in the `onMapReady` callback:

```
@Override
public void onMapReady(GoogleMap googleMap) {

mMap.setOnMarkerDragListener(this);
...
}
```

Then, update the `onMarkerDragEnd` method with the following code:

```
@Override
public void onMarkerDragEnd(Marker marker) {

    Memory memory = mMemories.get(marker.getId());
    updateMemoryPosition(memory, marker.getPosition());
    mDataSource.updateMemory(memory);

}
```

The previous code snippet updates the location when the marker is dragged.

InfoWindow click event

When a user clicks on `InfoWindow`, it allows him to delete the marker. To listen to the click events of `Infowindow`, we need to implement `GoogleMap.OnInfoWindowClickListener` and its callback method `onInfoWindowClick(..)`.
Register `infoWindoClicklistener` in the `onMapready` callback as follows:

```
mMap.setOnInfoWindowClickListener(this);
```

Inside the callback method, let's design an alert dialog when a user clicks on it. It should allow the user to delete the marker:

```
@Override
public void onInfoWindowClick(final Marker marker) {
    final Memory memory = mMemories.get(marker.getId());
    String[] actions  = {"Delete"};
```

```
        AlertDialog.Builder builder = new AlertDialog.Builder(this);
        builder.setTitle(memory.city+", "+memory.country)
                .setItems(actions, new DialogInterface.OnClickListener() {
                    @Override
                    public void onClick
                    (DialogInterface dialog, int which) {
                        if (which == 0){
                            marker.remove();
                            mDataSource.deleteMemory(memory);
                        }
                    }
                });

        builder.create().show();
    }
```

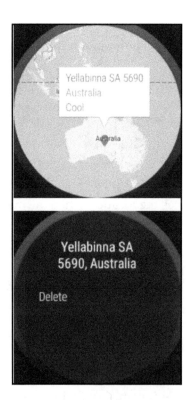

The UI controls

The UI controls such as zoom and location controls, are disabled for wearable devices. We can enable them using the `UISettings` class. The `UISettings` class extends to the object settings for the user interface of a Google Map. To obtain this interface, call `getUiSettings()`.

The following Boolean methods return the status of the component, whether it is enabled or disabled:

- `public boolean isCompassEnabled ()` : Gets whether the compass is enabled/disabled
- `public boolean isMyLocationButtonEnabled ()` : Gets whether the my-location button is enabled/disabled
- `public boolean isZoomControlsEnabled ()` : Gets whether the zoom controls are enabled/disabled
- `public boolean isZoomGesturesEnabled ()` : Gets whether zoom gestures are enabled/disabled
- `public boolean isTiltGesturesEnabled ()` : Gets whether tilt gestures are enabled/disabled
- `public boolean isRotateGesturesEnabled ()` : Gets whether rotate gestures are enabled/disabled
- `public boolean isScrollGesturesEnabled ()` : Gets whether scroll gestures are enabled/disabled

  These methods will return the status of the components.

  To enable these components for the application, `getUiSettings()` will provide appropriate set methods as follows:

- `public void setCompassEnabled (boolean enabled)`: Enables or disables the compass
- `public void setIndoorLevelPickerEnabled (boolean enabled)`: Sets whether the indoor level picker is enabled when the indoor mode is enabled
- `public void setMyLocationButtonEnabled (boolean enabled)`: Enables or disables the my-location button

- `public void setRotateGesturesEnabled (boolean enabled)`: Sets the preference for whether rotate gestures should be enabled or disabled
- `public void setZoomControlsEnabled (boolean enabled)`: Enables or disables the zoom controls

Let's see this in the `WearMapdiary` application. Let's enable the zoom controls for the application. In the `OnMapready` method, add the following line of code for the `mMap` object:

```
mMap.getUiSettings().setZoomControlsEnabled(true);
```

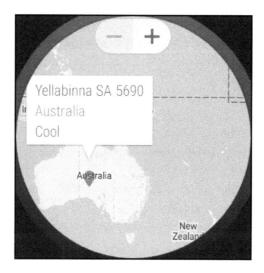

Similarly, we can set the other UI controls. There are certain limitations to all these controls on wearable devices, for example, `setIndoorLevelPickerEnabled` will not work on wearable devices.

Types of Maps

A map type governs the overall representation of a map. For example, an atlas usually contains political maps that focus on showing boundaries, and road maps show all of the roads for a city or a region. The Google Maps Android API offers four types of maps, as well as an option to have no map at all. Let's look at the options in more detail:

- **Normal**: Typical road map. Shows roads, some features built by humans, and important natural features like rivers. Road and feature labels are also visible.
- **Hybrid:** Satellite photograph data with road maps added. Road and feature labels are also visible.
- **Satellite:** Satellite photograph data. Road and feature labels are not visible.
- **Terrain:** Topographic data. The map includes colors, contour lines and labels, and perspective shading. Some roads and labels are also visible.
- **None:** No tiles. The map will be rendered as an empty grid with no tiles loaded.

Let's see this in the `WearMapdiary` application. Let's change the map type for the application. In the `OnMapready` method, add the following line of code in the `mMap` object to change the map type to Hybrid:

```
mMap.setMapType(GoogleMap.MAP_TYPE_HYBRID);
```

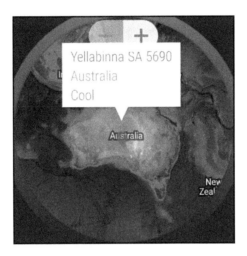

The Map is of a Hybrid type, which is a satellite image and labels. Now, to change the map type to Terrain, please insert the following code in mMap:

```
object'mMap.setMapType(GoogleMap.MAP_TYPE_TERRAIN);
```

The Terrain type map looks as shown preceding figure. Now, to change the map type to NONE, please insert the following code in the mMap object:

```
mMap.setMapType(GoogleMap.MAP_TYPE_NONE);
```

When you choose to have no maps, it looks as shown in the previous screenshot. Now, to change the map type to Satellite, please insert the following code in the mMap object:

```
mMap.setMapType(GoogleMap.MAP_TYPE_SATELLITE)
```

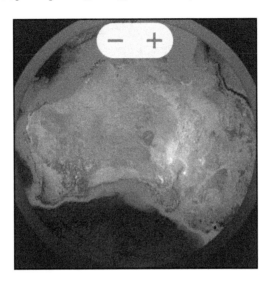

Streetview in Wear application

Google Street View provides panoramic 360-degree views from designated roads throughout its coverage area. Streetview is a great way for visualizing a user's destination or any address. Adding Streetview adds real-world visual elements to the application and provides a meaningful context to the users. Users can interact with Streetview; users will like to pan and scan the location in Streetview.

To create a street view map, we shall create a new fragment or activity and we can start the activity or attach the fragment. In this example, let's create a new activity with the SupportStreetViewPanoramaFragment class and launch the activity onMapLongclick callback:

```
Lets create a new activity with the following layout and java code.
//Java class
public class StreetView extends AppCompatActivity {

    private static final LatLng SYDNEY = new LatLng(-33.87365,
    151.20689);
```

```java
@Override
protected void onCreate(final Bundle savedInstanceState) {
    super.onCreate(savedInstanceState);
    setContentView(R.layout.activity_street_view);

    SupportStreetViewPanoramaFragment streetViewPanoramaFragment =
            (SupportStreetViewPanoramaFragment)
                    getSupportFragmentManager()
                    .findFragmentById(R.id.Streetviewpanorama);

    streetViewPanoramaFragment.getStreetViewPanoramaAsync(
            new OnStreetViewPanoramaReadyCallback() {
                @Override
                public void onStreetViewPanoramaReady
                (StreetViewPanorama panorama) {
                    // Only set the panorama to SYDNEY on startup
                    (when no panoramas have been
                    // loaded which is when the savedInstanceState
                    is null).
                    if (savedInstanceState == null) {
                        panorama.setPosition(SYDNEY);
                    }
                }
            });
}
}
```

Add the following code in the new layout resource and name the file as `activity_street_view`:

```xml
<?xml version="1.0" encoding="utf-8"?>

<FrameLayout xmlns:android="http://schemas.android.com/apk/res/android"
 android:layout_width="match_parent"
 android:layout_height="match_parent">

<fragment
 android:id="@+id/Streetviewpanorama"
 android:layout_width="match_parent"
 android:layout_height="match_parent"
 class="com.google.android.gms.maps.SupportStreetViewPanoramaFragment" />
</FrameLayout>
```

Now, start this activity `onMapLongclicklistner` in `MapActivity`. Before starting the activity, make sure you have changed the Application or Activity theme to `Theme.AppCompat.Light.NoActionBar`:

```
android:theme="@style/Theme.AppCompat.Light.NoActionBar"
@Override
    public void onMapLongClick(LatLng latLng) {
        // Display the dismiss overlay with a button to exit this
        activity.

        //          mDismissOverlay.show();

        Intent street = new Intent(MapsActivity.this,
        StreetView.class);
        startActivity(street);
```

Now, we have a complete, working, basic Streetview wear application, you can pan and rotate 360 degrees.

- **Polylines**: Polylines extend to the object class. A polyline is a list of points, where line segments are drawn between consecutive points. Polylines have the following properties:
    - **Points:** The vertices of the line. Line segments are drawn between consecutive points. The polyline needs start and end points for drawing the lines.

- **Width:** Line segment width in screen pixels. Width is a constant, independent of camera zoom.
- **Color:** Line segment color in ARGB format; the same format used by Color.
- **Start/End Cap:** Defines the shape to be used at the start or end of a polyline. Supported cap types: **ButtCap, SquareCap, RoundCap** (applicable for solid stroke pattern), and **CustomCap** (applicable for any stroke pattern). Default for both start and end is **ButtCap.**
- **Joint type:** The joint type defines the shape to be used when joining adjacent line segments at all the vertices of the polyline except the start and end vertices.
- **Stroke Pattern:** Solid or sequence of patternItem to be repeated along the line. Choices include the following:
  - Gap
  - Dash
  - Dot
- **Z-Index:** The order in which this tile overlay is drawn with respect to other overlays.
- **Visibility:** Indicates the line visibility, or tells whether the line is drawn or not.
- **Geodesic status:** Indicates whether the segments of the polyline should be drawn as geodesics, as opposed to straight lines on the Mercator projection. A geodesic is the shortest path between two points on the Earth's surface. The geodesic curve is constructed assuming the Earth is a sphere.
- **Clickability:** When you want to fire an event when a user clicks on the polyline. It works with `GoogleMap.OnPolylineClickListener` registered through `setOnPolylineClickListener(GoogleMap.OnPolylineClick Listener)`.
- **Tag:** An object associated with the polyline. For example, the object can contain data about what the polyline represents. This is easier than storing a separate `Map<Polyline, Object>`. As another example, you can associate a String ID corresponding to the ID from a dataset.

Add the following code in the `onMapready` callback and attach this to the map instance:

```
Polyline line = mMap.addPolyline(new PolylineOptions()
    .add(new LatLng(-34, 151), new LatLng(-37, 74.0))
    .width(5)
    .color(Color.WHITE));
```

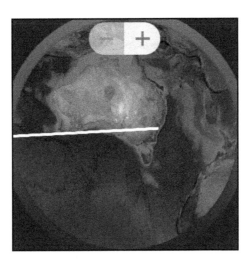

Best practices

Android Wear is very useful for quick and glanceable information. The most requested feature in wear is map in the latest Google Play services, updates for Google Maps came to Android Wear, which means you can develop the map application ideally, like how we develop for mobile applications there is no change in the process for developing wear map application. This means best-in-class development experience with just a few lines of code and configuration.

Let's talk about some common use cases for Android Wear maps applications and how to implement the best map application experience:

- One of the most common use cases is to simply display a map; since the wear has a small display, we might have to display the entire map in fullscreen.
Your application might need to show a marker to denote the landmark. Needs to allow users to pan around the map and find places on the map.

- Android Wear reserves the gesture of swiping from left to right for dismissing the current application. If you do not need your map to pan around, this will continue to work. However, if you need your map application to move and pan around the map, we need to override this particular dismiss gesture to reduce confusion and let the user exit the application. To do this, we can implement the `dismissoverlay` view. And attach it to a long click event. The view will handle the dismiss action.

- Another common use case is to select the location on the map, so you can share the location with your friends. To implement this, we can place the marker in the middle of the screen and let the user pan around the map and select the recent pan `latlong` value, which indicates the selected location within the map fragment component. Then, use the map `oncamerachange` listner to detect if the user has panned around the map. We can access the new location through the `cameraposition.target.letlong` value.

- It is a good practice to release components we are not using; for example, Google API clients when we initialize it. We shall release it in the activity lifecycle callbacks.

> For more information on implementing the best wear map application, follow this link: `https://developers.google.com/maps/documentation/android-api/wear`.

Summary

In this chapter, you have learned how to add UI controls like zoom, map types, and so on. Using the Google Maps Android API, you have learned the way in which users can interact with the wear map application with the following key items:

Adding the UI controls: UI controls help users to control the map in a more personalized manner.

Dragging the marker and updating location labels: When a user wants to modify the marker placement on the map, dragging the same marker is a great way.

Custom markers: We know that a marker identifies a location on a map. Customizing markers can help users to figure out the location type. Custom markers convey more about the location; for example, a fuel icon at the location conveys that the location is a fuel station.

Different map types: Different map types help users experience maps in a personalized manner.

Info window click event: An info window is a special kind of overlay for displaying content (normally, text or image) within a pop-up balloon at a given location on a map. `InfoWindow` click events help to do certain actions. For the scope of the WearMapDiary app, we are attaching `dialogfragment` for updating the text in the snippet area.

Polylines: A polyline specifies a series of coordinates as an array of LatLng objects. This represents a graphical path on the map.

Streetview: Google Street View provides panoramic 360-degree views from designated roads throughout its coverage area.

Now, with all these map-related ideas apart from `wearmapdiary`, we can produce the best of wear application that helps users.

8
Let us Chat in a Smart Way - Messaging API and More

The era of innovation has empowered us to chip away at numerous new shrewd subjects. Social media is currently an intense medium of communication. Taking a gander at the developing pattern of online networking and innovation, we could state that the belief system of social media has advanced and wiped out many difficulties of communication. Just about a couple of decades back, the communication medium was letters. A couple of centuries back, it was trained birds. If we still look back, we will definitely get a few more stories to comprehend the way people used to communicate in those days. Now, we are in the generation of IoT, wearable smart devices, and an era of smartphones, where communication happens across the planet in a fraction of a second. Without elaborating about communication, let's build a mobile and wear application that exhibits the power of Google wear messaging APIs to assist us in building chat application with a Wear companion application to administer and respond to the messages being received. To support the process of chatting, we will be using Google's very own technology Firebase in this chapter. We will not deep dive into the Firebase technologies, but we will surely understand the essentials of using Firebase in mobile platforms and working with wear technologies. A Firebase real-time database reflects the data being updated in its hashmap structure. Essentially, these are the stream of key-value pairs that Firebase works with. The data gets updated with minimal internet bandwidth requirement and instantly.

To support the process of chatting, we will be using Google's very own technology, Firebase, in this chapter. We will comprehend a generic registration and login process for the mobile platform and we will have space for all the registered members and enable every one of them to chat exclusively by picking one user from the list.

In this chapter, we will explore the following points:

- Configuring Firebase to your mobile application
- Creating user interface
- Working with `GoogleApiClient`
- Understanding the Message API
- Handling events
- Building a wear module

Now, let's understand how to set up Firebase to our project. It follows a few steps that we need to carry out before using Firebase technologies in the project. First, we need to apply the Firebase plugin and then the dependencies that we use in our project.

Installing Firebase

For installing Firebase, perform the following steps:

1. Visit the Firebase console `https://console.firebase.google.com`:

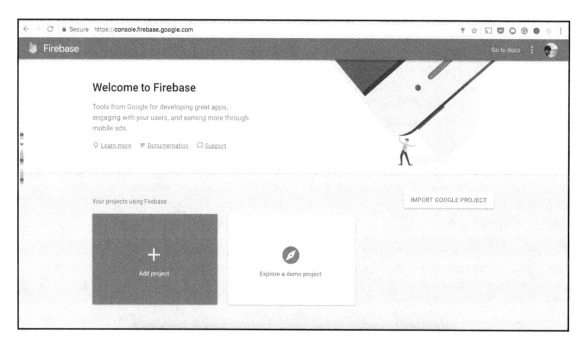

2. Choose **Add project** in the console and fill the necessary information about the project. After the project is successfully added, you will see the following screen:

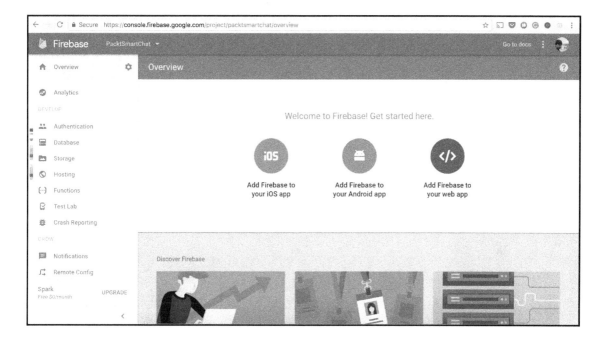

3. The get started page helps you set up the project for different platforms. Let's choose the second option, which says **Add Firebase to your Android app**:

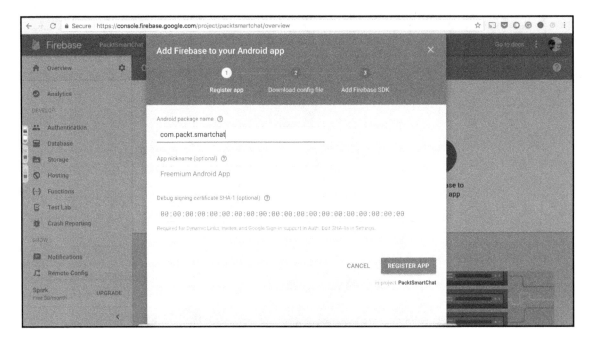

4. Add the project **package name** and, for further security purposes, you can add the **SHA-1** fingerprint, but that is optional. Now register the app:

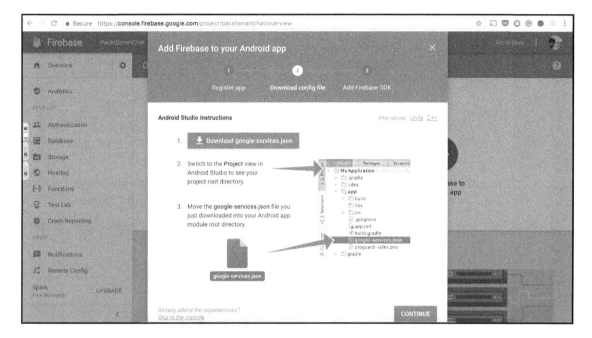

5. Download the config file. The `google-services.json` file will have all the important configuration for the app and place it in the app directory of your project structure.

Now, let's fire up Android Studio and create the project:

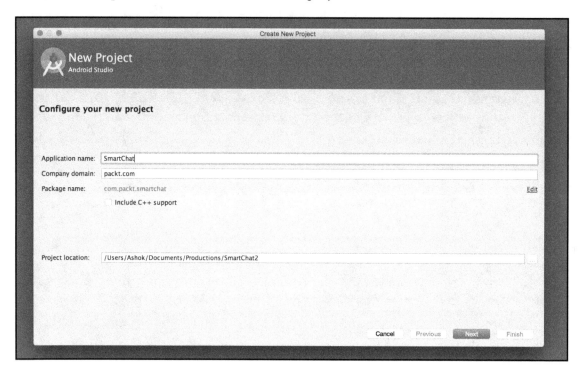

Make sure the package name is the same as the one mentioned in the Firebase console.

Let's choose the targeted platforms that are both phone and Wear:

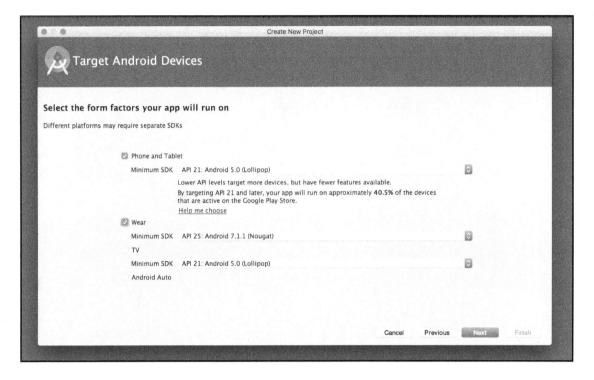

Now, add **Empty Activity** to the mobile activity chooser:

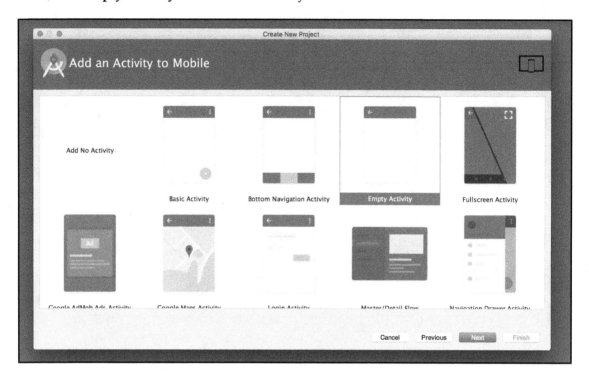

Select **Blank Wear Activity** in the wear activity chooser for generating blank wear activity code through the Android Studio template:

Now, name your class and XML files and finish the project for Android Studio to generate boiler for your mobile and wear module. Using file explorer or finder, go to the directory structure and copy and paste the `google-services.json` file:

Since we are building wear and mobile app together and the `app` directory name will be mobile for mobile and wear for wear projects, we shall copy the config file (`google-services.json`) inside the mobile directory.

After adding the config file, it is time to add the plugin classpath dependency:

```
classpath 'com.google.gms.google services:3.0.0'
```

Now, in the mobile Gradle module dependency, apply the plugin to the bottom of all the tag scope, as shown in the following screenshot:

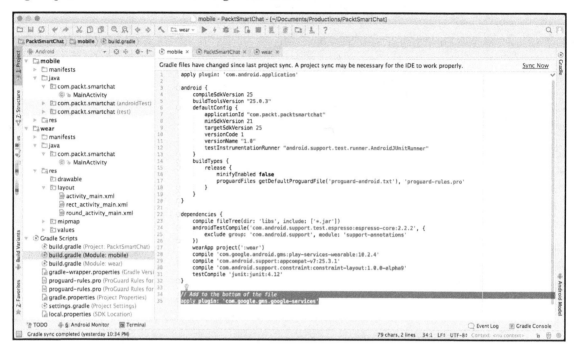

 To help Gradle to manage dependencies and the order of events that Gradle uses to build the project, we are supposed to add Google Play services dependencies at the bottom of the Gradle file. However, it will also avoid the conflict with other Google dependencies.

After a successful sync, the Firebase SDKs are integrated into our project. Now, we can get started using features that we are interested in. In this project, for the scope of a chatting feature, we will be using the Firebase Realtime database. Let's add the dependency to the same gradle file inside the dependencies. We will use the volley network library for fetching the user's list from the Firebase users node. We need to add the design support library for material design support:

```
compile 'com.firebase:firebase-client-android:2.5.2+'
compile 'com.android.volley:volley:1.0.0'
compile 'com.android.support:design:25.1.1'
compile 'com.android.support:cardview-v7:25.1.1'
```

On off the chance if you see gradle error please add the following packaging in gradle file under dependency section.

```
packagingOptions {
exclude 'META-INF/DEPENDENCIES.txt'
exclude 'META-INF/LICENSE.txt'
exclude 'META-INF/NOTICE.txt'
exclude 'META-INF/NOTICE'
exclude 'META-INF/LICENSE'
exclude 'META-INF/DEPENDENCIES'
exclude 'META-INF/notice.txt'
exclude 'META-INF/license.txt'
exclude 'META-INF/dependencies.txt'
exclude 'META-INF/LGPL2.1'
}
```

After all the necessary project setup, let's conceptualize the chatting application that we are going to build.

A basic chat application needs a registration process and, to avoid anonymous chats or, at least, to know with whom we are chatting, we need the sender and receiver names. The first screen is going to be the login screen with the username and password fields allowing already registered users to start chatting with the users. Then, we have the registration screen with the same username and password fields. Once the user successfully registers, we will ask the user to enter the credentials and allow them to have access to the list-of-users screen, where one can pick with whom they want to chat.

Conceptualizing the chatting application

A login screen with input fields for users to enter the credentials will look as follows:

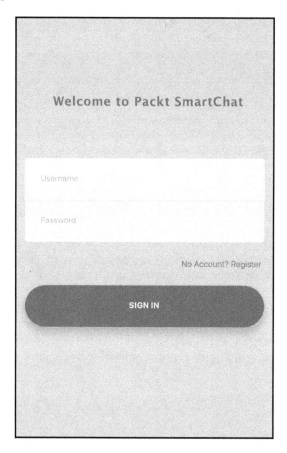

The registration screen with input fields will look as follows:

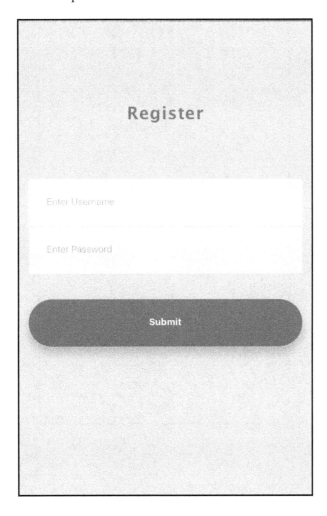

The following screenshot represents the user screen that shows the list of registered users:

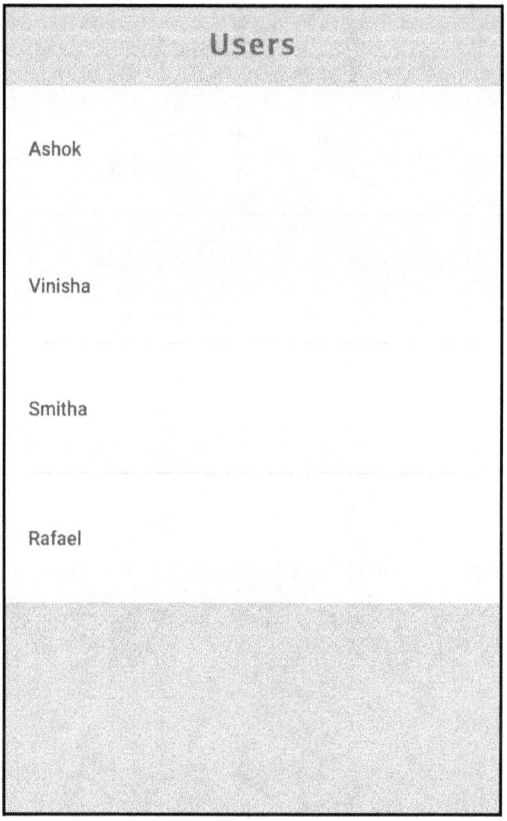

The chat screen with the actual chat messages will look as follows:

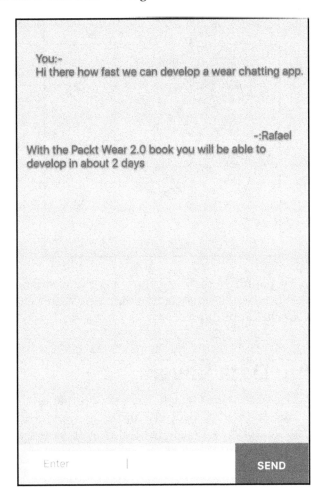

The Wear Chat Application will look as follows on round screens:

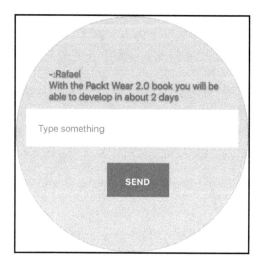

When a message enters the handheld device, it should notify wear and, from the wear device, users should be able to send a reply to that message. In this chapter, we will see a working mobile and wear chat application.

Understanding Data Layer

Wearable Data Layer API is part of the Google Play services that establishes the communication channel to handheld device apps and wear apps. Using the GoogleApiClient class, we can get access to the Data Layer. Primarily, the Data Layer is used in the Wear app to communicate with the handheld device, but using it for connecting to the network is discouraged. When we create the GoogleAPIClient class using the builder pattern, we will be attaching Wearable.API to the addAPI method. When we add multiple APIs in GoogleApiclient, there is a chance of the client instance failing into the onConnection fail callbacks. It's a good approach to add an API through addApiIfAvailable(). This will take care of most of the hard work; it will add the API if it's available. After adding all this using addConnectionCallbacks, we can handle the Data Layer events. We need to start the connection of the client instance by calling the connect() method. After a successful connection, we can use the Data Layer API.

Data Layer events

Events allow the developer to listen to what's happening in the communication channel. A successful communication channel will be able to send the status of the call when it's complete. These events will allow the developer to monitor all the state changes and data changes in a wireless communication channel. The Data Layer API returns pending results on an incomplete transaction, such as `putdataitem()`. The pending results will be automatically queued in the background when the transaction is incomplete and, if we don't handle it, this operation will be completed in the background. However, pending results need to be handled; pending result will wait for the result status and it has two methods to wait for the results synchronously and asynchronously.

If the Data Layer code is running in the UI thread, we should avoid making blocking calls to the Data Layer API. Using asynchronous callbacks to the `pendingresult` object, we will be able to check the status and other vital information:

```
pendingResult.setResultCallback(new ResultCallback<DataItemResult>() {
    @Override
    public void onResult(final DataItemResult result) {
        if(result.getStatus().isSuccess()) {
            Log.d(TAG, "Data item set: " +
            result.getDataItem().getUri());
        }
    }
});
```

If the Data Layer code is running in a separate thread in a background service, such as `wearableListenerService`, it's alright to block the calls and you can call the `await()` method on the `pendingresult` object:

```
DataItemResult result = pendingResult.await();
if(result.getStatus().isSuccess()) {
    Log.d(TAG, "Data item set: " + result.getDataItem().getUri());
}
```

The Data Layer events can be monitored in two ways:

- Creating a class that extends to `WearableListenerService`
- An activity that implements `DataApi.DataListener`

In both the facilities, we override the methods to handle the data events. Typically, we need to create the instance in both wearable and handheld apps. We can override the methods that we need for the application use case. Essentially, `WearableListenerService` has the following events:

- `onDataChanged()`: Whenever this is created, deleted, or updated, the system will trigger this method.
- `onMessageReceived()`: A message sent from a node triggers this event in the target node.
- `onCapabilityChanged()`: This event is triggered when a capability of an instance advertise becomes available on the network. We can check the nearby nodes by calling `isnearby()`.

These methods are executed in a background thread.

To create `WearableListenerService`, we need to create a class that extends `WearableListenerService`. Listen for the events that you're interested in, such as `onDataChanged()`. Declare an `intent` filter in your Android manifest to notify the system about your `WearableListenerService`:

```
public class DataLayerListenerService extends WearableListenerService {

    private static final String TAG = "DataLayerSample";
    private static final String START_ACTIVITY_PATH = "/start-
    activity";
    private static final String DATA_ITEM_RECEIVED_PATH = "/data-item-
    received";

    @Override
    public void onDataChanged(DataEventBuffer dataEvents) {
        if (Log.isLoggable(TAG, Log.DEBUG)) {
            Log.d(TAG, "onDataChanged: " + dataEvents);
        }

        GoogleApiClient googleApiClient = new
        GoogleApiClient.Builder(this)
                .addApi(Wearable.API)
                .build();

        ConnectionResult connectionResult =
        googleApiClient.blockingConnect(30, TimeUnit.SECONDS);

        if (!connectionResult.isSuccess()) {
            Log.e(TAG, "Failed to connect to GoogleApiClient.");
            return;
```

```
        }

        // Loop through the events and send a message
        // to the node that created the data item.
        for (DataEvent event : dataEvents) {
            Uri uri = event.getDataItem().getUri();

            // Get the node id from the host value of the URI
            String nodeId = uri.getHost();
            // Set the data of the message to be the bytes of the URI
            byte[] payload = uri.toString().getBytes();

            // Send the RPC
            Wearable.MessageApi.sendMessage(googleApiClient, nodeId,
            DATA_ITEM_RECEIVED_PATH, payload);
        }
    }
}
```

And register the service in the manifest as follows:

```
<service android:name=".DataLayerListenerService">
  <intent-filter>
      <action
      android:name="com.google.android.gms.wearable.DATA_CHANGED" />
      <data android:scheme="wear" android:host="*"
              android:path="/start-activity" />
  </intent-filter>
</service>
```

The DATA_CHANGED action replaces the previously recommended BIND_LISTENER action so that only specific events go through the path. We will understand more while we are working on a live project in this chapter.

Capability API

This API helps in advertising the capabilities given by nodes in the wear network. Capabilities are local to the application. Utilizing the Data Layer and Message API, we can communicate with the nodes. To discover whether the target node is proficient in doing certain actions, we have to utilize the capability API, for instance on the off chance that we need to launch an activity from wear.

To initialize the capability API to your application, perform the following steps:

1. Create an XML configuration file in the `res/values` directory.
2. Add a resource named `android_wear_capabilities`.
3. Define the capability that the device provides:

```
<resources>
    <string-array name="android_wear_capabilities">
        <item>voice_transcription</item>
    </string-array>
</resources>
```

Java program for `voice_transcription`:

```
private static final String
        VOICE_TRANSCRIPTION_CAPABILITY_NAME = "voice_transcription";

private GoogleApiClient mGoogleApiClient;

...

private void setupVoiceTranscription() {
    CapabilityApi.GetCapabilityResult result =
            Wearable.CapabilityApi.getCapability(
                    mGoogleApiClient,
                    VOICE_TRANSCRIPTION_CAPABILITY_NAME,
                    CapabilityApi.FILTER_REACHABLE).await();

    updateTranscriptionCapability(result.getCapability());
}
```

Now that we have all the setup and designs to implement the chat application, let's get started.

Mobile app implementation

The mobile app for the chat application utilises Google's Firebase real-time database. Whenever a user sends a message, it reflects in the Firebase console in real time. Stories aside, now that we have all the screens up and ready, let's get started on writing the code.

Since we have the color that we are going to be using, let's declare the colors in the colors value XML file under the `res` directory:

```
<?xml version="1.0" encoding="utf-8"?>
<resources>
```

```
<color name="colorPrimary">#129793</color>
<color name="colorPrimaryDark">#006865</color>
<color name="colorAccent">#ffaf10</color>
<color name="white">#fff</color>
</resources>
```

As per the design, we have a curved edge button with a teal color background. To make a similar button, we need to create an XML resource in the `drawable` directory and call it as `buttonbg.xml`, which is basically a `selector` tag, as follows:

```xml
<?xml version="1.0" encoding="utf-8"?>
<selector xmlns:android="http://schemas.android.com/apk/res/android">

    <!-- When item pressed this item will be triggered -->
    <item android:state_pressed="true">
        <shape android:shape="rectangle">
        <corners android:radius="25dp" />
        <solid android:color="@color/colorPrimaryDark" />
    </shape>
    </item>

    <!-- By default the background will be this item -->
    <item>
        <shape android:shape="rectangle">
        <corners android:radius="25dp" />
        <solid android:color="@color/colorPrimary" />
    </shape>
</item>

</selector>
```

Inside the `selector` tag, we have an `item` tag that conveys the states; any normal button will have states, such as clicked, released, and default. Here, we have taken the default button background and pressed state and, using the `item` property tags, such as shape and corners, we are carving the button, as shown in the design.

To the comfort of not having multiple changes, we will not refactor `MainActivity` into `LoginActivity`, rather we will consider `MainActivity` as the `LoginActivity`. Now, in `activity_main.xml`, let's add the following code to the login screen design. For making the screen dynamic, we will be adding the code under `scrollview`:

```xml
<?xml version="1.0" encoding="utf-8"?>
<ScrollView xmlns:android="http://schemas.android.com/apk/res/android"
    xmlns:app="http://schemas.android.com/apk/res-auto"
    xmlns:tools="http://schemas.android.com/tools"
    android:layout_width="match_parent"
    android:layout_height="match_parent"
```

```
        android:fillViewport="true"
        android:fitsSystemWindows="true">

    <RelativeLayout
        android:layout_width="match_parent"
        android:layout_height="match_parent">
      <!-- Your design code goes here -->

    </RelativeLayout>
</ScrollView>
```

Now, for completing the login design, we need two input fields, one button instance, and one clickable link instance. The completed login screen code would look as follows:

```
<?xml version="1.0" encoding="utf-8"?>
<ScrollView xmlns:android="http://schemas.android.com/apk/res/android"
    xmlns:app="http://schemas.android.com/apk/res-auto"
    xmlns:tools="http://schemas.android.com/tools"
    android:layout_width="match_parent"
    android:layout_height="match_parent"
    android:fillViewport="true"
    android:fitsSystemWindows="true">

    <RelativeLayout
        android:layout_width="match_parent"
        android:layout_height="match_parent">


        <LinearLayout
            android:layout_width="match_parent"
            android:layout_height="match_parent"
            android:layout_marginTop="25dp"
            android:orientation="vertical">

            <TextView
                android:layout_width="wrap_content"
                android:layout_height="wrap_content"
                android:layout_gravity="center_horizontal"
                android:layout_marginBottom="24dp"
                android:gravity="center|center_horizontal
                |center_vertical"
                android:text="Welcome to Packt Smartchat"
                android:textColor="@color/colorPrimaryDark"
                android:textSize="25sp"
                android:textStyle="bold" />

            <!-- Email Label -->
```

```
<android.support.v7.widget.CardView
    android:layout_margin="10dp"
    android:layout_width="match_parent"
    android:layout_height="wrap_content"
    android:orientation="vertical">

    <LinearLayout
        android:padding="5dp"
        android:orientation="vertical"
        android:layout_width="match_parent"
        android:layout_height="wrap_content">

    <android.support.design.widget.TextInputLayout
        android:layout_width="match_parent"
        android:layout_height="wrap_content"
        android:layout_marginBottom="8dp"
        android:layout_marginTop="8dp">

        <EditText
            android:id="@+id/input_email"
            android:layout_width="match_parent"
            android:layout_height="wrap_content"
            android:hint="Username"
            android:inputType="textEmailAddress"
            android:windowSoftInputMode="stateHidden" />
    </android.support.design.widget.TextInputLayout>

    <!-- Password Label -->
    <android.support.design.widget.TextInputLayout
        android:layout_width="match_parent"
        android:layout_height="wrap_content"
        android:layout_marginBottom="8dp"
        android:layout_marginTop="8dp"
        app:passwordToggleEnabled="true">

        <EditText
            android:id="@+id/input_password"
            android:layout_width="match_parent"
            android:layout_height="wrap_content"
            android:hint="Password"
            android:inputType="textPassword"
            android:windowSoftInputMode="stateHidden" />
    </android.support.design.widget.TextInputLayout>
    </LinearLayout>
</android.support.v7.widget.CardView>


<LinearLayout
```

```
            android:layout_width="fill_parent"
            android:layout_height="wrap_content"
            android:orientation="horizontal"
            android:paddingBottom="5dp"
            android:paddingTop="8dp">

            <TextView
                android:id="@+id/register"
                android:layout_width="wrap_content"
                android:layout_height="wrap_content"
                android:layout_weight="1"
                android:gravity="end"
                android:padding="5dp"
                android:text="No Account? Register"
                android:textSize="14sp" />
        </LinearLayout>

        <Button
            android:id="@+id/btn_login"
            android:layout_width="150dp"
            android:layout_height="wrap_content"
            android:layout_gravity="center"
            android:layout_marginBottom="24dp"
            android:layout_marginTop="24dp"
            android:background="@drawable/buttonbg"
            android:textColor="@color/white"
            android:textStyle="bold"
            android:padding="12dp"
            android:text="Login" />


    </LinearLayout>
  </RelativeLayout>

</ScrollView>
```

Let's create another activity and call it `RegistrationActivity`, which has a similar component requirement as the login activity, two input fields, and one button. The complete code for the XML layout looks as follows:

```
<?xml version="1.0" encoding="utf-8"?>
<ScrollView xmlns:android="http://schemas.android.com/apk/res/android"
    xmlns:app="http://schemas.android.com/apk/res-auto"
    xmlns:tools="http://schemas.android.com/tools"
    android:layout_width="match_parent"
    android:layout_height="match_parent"
    android:fillViewport="true"
    android:fitsSystemWindows="true">
```

```
<RelativeLayout
    android:layout_width="match_parent"
    android:layout_height="match_parent">


<LinearLayout
    android:layout_width="match_parent"
    android:layout_height="match_parent"
    android:layout_marginTop="25dp"
    android:orientation="vertical">

<TextView
    android:layout_width="wrap_content"
    android:layout_height="wrap_content"
    android:layout_gravity="center_horizontal"
    android:layout_marginBottom="24dp"
    android:gravity="center
    |center_horizontal|center_vertical"
    android:text="Register"
    android:textColor="@color/colorPrimaryDark"
    android:textSize="25sp"
    android:textStyle="bold" />

<!--  Email Label -->

<android.support.v7.widget.CardView
    android:layout_margin="10dp"
    android:layout_width="match_parent"
    android:layout_height="wrap_content"
    android:orientation="vertical">

<LinearLayout
    android:padding="5dp"
    android:orientation="vertical"
    android:layout_width="match_parent"
    android:layout_height="wrap_content">

<android.support.design.widget.TextInputLayout
    android:layout_width="match_parent"
    android:layout_height="wrap_content"
    android:layout_marginBottom="8dp"
    android:layout_marginTop="8dp">

<EditText
    android:id="@+id/input_email"
    android:layout_width="match_parent"
    android:layout_height="wrap_content"
    android:hint="Username"
```

```
                android:inputType="textEmailAddress"
                android:windowSoftInputMode="stateHidden"
            />
        </android.support.design.widget.TextInputLayout>

        <!-- Password Label -->
        <android.support.design.widget.TextInputLayout
            android:layout_width="match_parent"
            android:layout_height="wrap_content"
            android:layout_marginBottom="8dp"
            android:layout_marginTop="8dp"
            app:passwordToggleEnabled="true">

            <EditText
                android:id="@+id/input_password"
                android:layout_width="match_parent"
                android:layout_height="wrap_content"
                android:hint="Password"
                android:inputType="textPassword"
                android:windowSoftInputMode="stateHidden"
            />
        </android.support.design.widget.TextInputLayout>
    </LinearLayout>
</android.support.v7.widget.CardView>


<LinearLayout
    android:layout_width="fill_parent"
    android:layout_height="wrap_content"
    android:orientation="horizontal"
    android:paddingBottom="5dp"
    android:paddingTop="8dp">

</LinearLayout>

<Button
    android:id="@+id/btn_submit"
    android:layout_width="150dp"
    android:layout_height="wrap_content"
    android:layout_gravity="center"
    android:layout_marginBottom="24dp"
    android:layout_marginTop="24dp"
    android:background="@drawable/buttonbg"
    android:textColor="@color/white"
    android:textStyle="bold"
    android:padding="12dp"
    android:text="Submit" />
```

```
        </LinearLayout>
    </RelativeLayout>

  </ScrollView>
```

Now, let's create a list of the user's activity that will have a list of users. Call it as the `UsersList` activity and it will have one simple `ListView` and `TextView` for addressing the empty list. The complete XML code for `UsersListActivity` will look as follows:

```
<?xml version="1.0" encoding="utf-8"?>
<LinearLayout xmlns:android="http://schemas.android.com/apk/res/android"
    xmlns:tools="http://schemas.android.com/tools"
    android:layout_width="match_parent"
    android:layout_height="match_parent"
    android:orientation="vertical"
    android:padding="16dp">

    <TextView
        android:id="@+id/noUsersText"
        android:layout_width="match_parent"
        android:layout_height="wrap_content"
        android:text="No users found!"
        android:visibility="gone" />

    <ListView
        android:id="@+id/usersList"
        android:layout_width="match_parent"
        android:layout_height="wrap_content" />

</LinearLayout>
```

Create another activity for the chat screen. We will call it `ChatActivity`. Add the following code inside the activity XML file:

```
<?xml version="1.0" encoding="utf-8"?>
<LinearLayout xmlns:android="http://schemas.android.com/apk/res/android"
    xmlns:tools="http://schemas.android.com/tools"
    android:layout_width="match_parent"
    android:layout_height="match_parent"
    android:background="#ffffff"
    android:orientation="vertical"
    tools:context="com.packt.smartchat.MainActivity">

    <ScrollView
        android:layout_width="match_parent"
        android:layout_weight="20"
        android:layout_height="wrap_content"
        android:id="@+id/scrollView">
```

```xml
        <LinearLayout
            android:layout_width="match_parent"
            android:layout_height="wrap_content"
            android:orientation="vertical"
            android:id="@+id/layout1">
        </LinearLayout>

    </ScrollView>

    <include
        layout="@layout/message_area"
        android:layout_width="match_parent"
        android:layout_height="wrap_content"
        android:gravity="bottom"
        android:layout_marginTop="5dp"/>
</LinearLayout>
```

We need to include a layout for editing the message. Create another XML file named `message_area.xml` inside the `layout` directory and add the following code:

```xml
<?xml version="1.0" encoding="utf-8"?>
<LinearLayout xmlns:android="http://schemas.android.com/apk/res/android"
    android:layout_width="match_parent"
    android:layout_height="match_parent"
    android:background="@color/colorPrimaryDark"
    android:gravity="bottom"
    android:orientation="horizontal">

    <EditText
        android:layout_width="match_parent"
        android:layout_height="wrap_content"
        android:layout_weight="1"
        android:textColorHint="#CFD8DC"
        android:textColor="#CFD8DC"
        android:singleLine="true"
        android:hint="Write a message..."
        android:id="@+id/messageArea"
        android:maxHeight="80dp" />

    <ImageView
        android:layout_width="match_parent"
        android:layout_height="wrap_content"
        android:layout_weight="4"
        android:padding="4dp"
        android:src="@android:drawable/ic_menu_send"
        android:id="@+id/sendButton"/>
</LinearLayout>
```

Now, all our visual elements are in place to get started on writing our programming logic.

Add this following permission in the manifest file before we start working on our activity Java files:

```
<uses-permission android:name="android.permission.WAKE_LOCK" />
<uses-permission android:name="android.permission.INTERNET" />
```

In the `MainActivity` file, let's create all the instances and map them to their XML ID that we placed in `activity_main.xml`. Inside the `MainActivity` class, in the global scope, declare the following instances:

```
private TextView mRegister;
private EditText mUsername, mPassword;
private Button mLoginButton;
public String mUserStr, mPassStr;
```

Now, let's connect all these instances to their XML visual elements, as follows, using the `findViewById()` method inside the `oncreate` method:

```
mRegister = (TextView)findViewById(R.id.register);
mUsername = (EditText)findViewById(R.id.input_email);
mPassword = (EditText)findViewById(R.id.input_password);
mLoginButton = (Button)findViewById(R.id.btn_login);
```

Now, when the user clicks on the **Register** link, it should take the user to the registration activity. Using `intent`, we will achieve it:

```
mRegister.setOnClickListener(new View.OnClickListener() {

    @Override
    public void onClick(View v) {
        startActivity(new Intent(MainActivity.this,
        RegistrationActivity.class));
    }
});
```

By clicking on the **login** button, it should do a network call, check whether the user exists in Firebase, and show the proper action on success. Before we write the login logic, let's write the registration logic.

In registration, activity connects all the components in the Java file using the `findViewById()` method:

```
//In Global scope of registration activity
private EditText mUsername, mPassword;
private Button mSubmitButton;
```

```
public String mUserStr, mPassStr;

// Inside the oncreate method
Firebase.setAndroidContext(this);
mUsername = (EditText)findViewById(R.id.input_email);
mPassword = (EditText)findViewById(R.id.input_password);
mSubmitButton = (Button)findViewById(R.id.btn_submit);
```

Attach a click listener to `mSubmit` and fetch the inputs in the `onClick` listener to make sure we are not passing an empty string:

```
mSubmitButton.setOnClickListener(new View.OnClickListener() {
    @Override
    public void onClick(View v) {
      // Input fields
      // Validation logics
    }
});
```

Having a simple validation check will make the app strong from error prone situations. The validation and fetching input from the input fields are as follows:

```
mSubmitButton.setOnClickListener(new View.OnClickListener() {
    @Override
    public void onClick(View v) {
        mUserStr = mUsername.getText().toString();
        mPassStr = mPassword.getText().toString();

        // Validation
        if(mUserStr.equals("")){
            mUsername.setError("can't be blank");
        }
        else if(mPassStr.equals("")){
            mPassword.setError("can't be blank");
        }
        else if(!mUserStr.matches("[A-Za-z0-9]+")){
            mUsername.setError("only alphabet or number allowed");
        }
        else if(mUserStr.length()<5){
            mUsername.setError("at least 5 characters long");
        }
        else if(mPassStr.length()<5){
            mPassword.setError("at least 5 characters long");
        }
    }
});
```

We now need to reach Firebase for registering the user. Before we proceed, let's log in to the Firebase console, `https://console.firebase.google.com`, and go to the project that we created before.

Now, on the left-hand side menu, we will see the database option and choose it. In the rules tab, by default, the read and write authorizations are set to `null`. It would be ideal if you change it to `true`, but this is not suggested when you are writing a production application:

```
{
  "rules": {
  ".read": true,
  ".write": true
  }
}
```

 When we set a read and write permission to true, essentially, we are telling Firebase anyone can read and write on the off chance that they have an endpoint URL.

Knowing the intricacies of making the URL public, we will use it in the project. Now, in the `mSubmit` click listener, we will be checking for a few validations and fetch the username and password.

We should finish the code for the `mSubmit` click listener. After the `else if` instance of the password key field, let's make an else case for doing all the Firebase network operations. We will make the Firebase reference URL, push the child values, and utilize the `volley` network library. We will check whether the username exists and, on the off chance that it exists, we will allow the user to use the application.

The Firebase endpoint URL for this project is `https://packt-wear.firebaseio.com` and the node name can be anything we wish to add for users. Let's add `https://packt-wear.firebaseio.com/users`; the code looks as follows:

```
else {
        final ProgressDialog pd = new
        ProgressDialog(RegistrationActivity.this);
        pd.setMessage("Loading...");
        pd.show();

        String url = "https://packt-wear.firebaseio.com/users.json";

        StringRequest request = new StringRequest(Request.Method.GET,
        url, new Response.Listener<String>(){
            @Override
```

```
public void onResponse(String s) {
    Firebase reference = new Firebase("https://packt-
    wear.firebaseio.com/users");
    if(s.equals("null")) {
        reference.child(mUserStr)
        .child("password").setValue(mPassStr);
        Toast.makeText(RegistrationActivity.this,
        "registration successful",
        Toast.LENGTH_LONG).show();
    }
    else {
        try {
            JSONObject obj = new JSONObject(s);

            if (!obj.has(mUserStr)) {
                reference.child(mUserStr)
                .child("password").setValue(mPassStr);
                Toast.makeText(RegistrationActivity.this,
                "registration successful",
                Toast.LENGTH_LONG).show();
            } else {
                Toast.makeText(RegistrationActivity.this,
                "username already exists",
                Toast.LENGTH_LONG).show();
            }

        } catch (JSONException e) {
            e.printStackTrace();
        }
    }

    pd.dismiss();
}

},new Response.ErrorListener(){
    @Override
    public void onErrorResponse(VolleyError volleyError) {
        System.out.println("" + volleyError );
        pd.dismiss();
    }
});

RequestQueue rQueue =
Volley.newRequestQueue(RegistrationActivity.this);
rQueue.add(request);
    }
}
```

Using `volley`, we can add the request queues and handle the network request in a very efficient manner.

Now, the complete registration activity class looks as follows:

```
public class RegistrationActivity extends AppCompatActivity {

    private EditText mUsername, mPassword;
    private Button mSubmitButton;
    public String mUserStr, mPassStr;

    @Override
    protected void onCreate(Bundle savedInstanceState) {
        super.onCreate(savedInstanceState);
        setContentView(R.layout.activity_registration);

        mUsername = (EditText)findViewById(R.id.input_email);
        mPassword = (EditText)findViewById(R.id.input_password);
        mSubmitButton = (Button)findViewById(R.id.btn_submit);

        Firebase.setAndroidContext(this);

        mSubmitButton.setOnClickListener(new View.OnClickListener() {
            @Override
            public void onClick(View v) {
                mUserStr = mUsername.getText().toString();
                mPassStr = mPassword.getText().toString();

                // Validation
                if(mUserStr.equals("")){
                    mUsername.setError("can't be blank");
                }
                else if(mPassStr.equals("")){
                    mPassword.setError("can't be blank");
                }
                else if(!mUserStr.matches("[A-Za-z0-9]+")){
                    mUsername.setError("only alphabet or number
                    allowed");
                }
                else if(mUserStr.length()<5){
                    mUsername.setError("at least 5 characters long");
                }
                else if(mPassStr.length()<5){
                    mPassword.setError("at least 5 characters long");
                }else {
                    final ProgressDialog pd = new
                    ProgressDialog(RegistrationActivity.this);
                    pd.setMessage("Loading...");
```

```
pd.show();

String url = "https://packt-
wear.firebaseio.com/users.json";

StringRequest request = new StringRequest
(Request.Method.GET, url,
new Response.Listener<String>(){
    @Override
    public void onResponse(String s) {
        Firebase reference = new Firebase
    ("https://packt-wear.firebaseio.com/users");
        if(s.equals("null")) {
            reference.child(mUserStr)
            .child("password").setValue(mPassStr);
            Toast.makeText
            (RegistrationActivity.this,
            "registration successful",
            Toast.LENGTH_LONG).show();
        }
        else {
            try {
                JSONObject obj = new JSONObject(s);

                if (!obj.has(mUserStr)) {
                    reference.child(mUserStr)
                    .child("password")
                    .setValue(mPassStr);
                    Toast.makeText
                    (RegistrationActivity.this,
                    "registration successful",
                    Toast.LENGTH_LONG).show();
                } else {
                    Toast.makeText
                    (RegistrationActivity.this,
                    "username already exists",
                    Toast.LENGTH_LONG).show();
                }

            } catch (JSONException e) {
                e.printStackTrace();
            }
        }

        pd.dismiss();
    }

},new Response.ErrorListener(){
```

```
                @Override
                public void onErrorResponse(VolleyError
                volleyError) {
                    System.out.println("" + volleyError );
                    pd.dismiss();
                }
            });

            RequestQueue rQueue =
            Volley.newRequestQueue
            (RegistrationActivity.this);
            rQueue.add(request);
        }
      }
    });

  }
}
```

Now, let's jump into `MainActivity` for user login logic.

Before we continue, let's create a class with static instances as follows:

```
public class User {
    static String username = "";
    static String password = "";
    static String chatWith = "";
}
```

Now, as we have seen in the registration screen, let's validate it inside the login screen using the `volley` library, let's check whether the username exists. If a valid user with a valid password logs in, we will have to allow the user to the chat screen. The following code goes inside the login click listener:

```
mUserStr = mUsername.getText().toString();
mPassStr = mPassword.getText().toString();

if(mUserStr.equals("")){
    mUsername.setError("Please enter your username");
}
else if(mPassStr.equals("")){
    mPassword.setError("can't be blank");
}
else{
    String url = "https://packt-wear.firebaseio.com/users.json";
    final ProgressDialog pd = new
    ProgressDialog(MainActivity.this);
    pd.setMessage("Loading...");
```

```
pd.show();

StringRequest request = new StringRequest(Request.Method.GET,
url, new Response.Listener<String>(){
    @Override
    public void onResponse(String s) {
        if(s.equals("null")){
            Toast.makeText(MainActivity.this, "user not found",
            Toast.LENGTH_LONG).show();
        }
        else{
            try {
                JSONObject obj = new JSONObject(s);

                if(!obj.has(mUserStr)){
                    Toast.makeText(MainActivity.this, "user not
                    found", Toast.LENGTH_LONG).show();
                }
                else if(obj.getJSONObject(mUserStr)
                .getString("password").equals(mPassStr)){
                    User.username = mUserStr;
                    User.password = mPassStr;
                    startActivity(new Intent(MainActivity.this,
                    UsersListActivity.class));
                }
                else {
                    Toast.makeText(MainActivity.this,
                    "incorrect password", Toast
                    .LENGTH_LONG).show();
                }
            } catch (JSONException e) {
                e.printStackTrace();
            }
        }

        pd.dismiss();
    }
},new Response.ErrorListener(){
    @Override
    public void onErrorResponse(VolleyError volleyError) {
        System.out.println("" + volleyError);
        pd.dismiss();
    }
});

RequestQueue rQueue =
Volley.newRequestQueue(MainActivity.this);
rQueue.add(request);
```

```
        }
    }
```

The complete class will look as follows:

```java
public class MainActivity extends AppCompatActivity {

    private TextView mRegister;
    private EditText mUsername, mPassword;
    private Button mLoginButton;
    public String mUserStr, mPassStr;

    @Override
    protected void onCreate(Bundle savedInstanceState) {
        super.onCreate(savedInstanceState);
        setContentView(R.layout.activity_main);

        mRegister = (TextView)findViewById(R.id.register);
        mUsername = (EditText)findViewById(R.id.input_email);
        mPassword = (EditText)findViewById(R.id.input_password);
        mLoginButton = (Button)findViewById(R.id.btn_login);

        mRegister.setOnClickListener(new View.OnClickListener() {

            @Override
            public void onClick(View v) {
                startActivity(new Intent(MainActivity.this,
                RegistrationActivity.class));
            }
        });

        mLoginButton.setOnClickListener(new View.OnClickListener() {
            @Override
            public void onClick(View v) {
                mUserStr = mUsername.getText().toString();
                mPassStr = mPassword.getText().toString();

                if(mUserStr.equals("")){
                    mUsername.setError("Please enter your username");
                }
                else if(mPassStr.equals("")){
                    mPassword.setError("can't be blank");
                }
                else{
                    String url = "https://packt-
                    wear.firebaseio.com/users.json";
                    final ProgressDialog pd = new
                    ProgressDialog(MainActivity.this);
```

```java
pd.setMessage("Loading...");
pd.show();

StringRequest request = new StringRequest
(Request.Method.GET, url,
new Response.Listener<String>(){
    @Override
    public void onResponse(String s) {
        if(s.equals("null")){
            Toast.makeText(MainActivity.this, "user
            not found", Toast.LENGTH_LONG).show();
        }
        else{
            try {
                JSONObject obj = new JSONObject(s);

                if(!obj.has(mUserStr)){
                Toast.makeText(MainActivity.this,
                "user not found",
                Toast.LENGTH_LONG).show();
                }
                else if(obj.getJSONObject(mUserStr)
                .getString("password")
                .equals(mPassStr)){
                    User.username = mUserStr;
                    User.password = mPassStr;
                    startActivity(new
                    Intent(MainActivity.this,
                    UsersListActivity.class));
                }
                else {
                Toast.makeText(MainActivity.this,
                "incorrect password",
                Toast.LENGTH_LONG).show();
                }
            } catch (JSONException e) {
                e.printStackTrace();
            }
        }

        pd.dismiss();
    }
},new Response.ErrorListener(){
    @Override
    public void onErrorResponse(VolleyError
    volleyError) {
        System.out.println("" + volleyError);
        pd.dismiss();
```

```
                            }
                        });

                        RequestQueue rQueue =
                        Volley.newRequestQueue(MainActivity.this);
                        rQueue.add(request);
                    }
                }
            });

        }
    }
```

Now, after allowing the user to have a successful login, we need to show a list of users, ignoring the one who logged in. But the user should be able to see other lists of users. Now, let's work on getting the list of users in ListView. Let's connect the components:

```
//Instances
private ListView mUsersList;
private TextView mNoUsersText;
private ArrayList<String> mArraylist = new ArrayList<>();
private int totalUsers = 0;
private ProgressDialog mProgressDialog;

//inside onCreate method
mUsersList = (ListView)findViewById(R.id.usersList);
mNoUsersText = (TextView)findViewById(R.id.noUsersText);
mProgressDialog = new ProgressDialog(UsersListActivity.this);

mProgressDialog.setMessage("Loading...");
mProgressDialog.show();
```

Now, right in the onCreate method, we will initiate the volley and fetch the list of users, as follows:

```
String url = "https://packt-wear.firebaseio.com/users.json";
    StringRequest request = new StringRequest(Request.Method.GET, url,
    new Response.Listener<String>(){
        @Override
        public void onResponse(String s) {
            doOnSuccess(s);
        }
    },new Response.ErrorListener(){
        @Override
        public void onErrorResponse(VolleyError volleyError) {
            System.out.println("" + volleyError);
        }
    });
```

```
        RequestQueue rQueue =
        Volley.newRequestQueue(UsersListActivity.this);
        rQueue.add(request);

        mUsersList.setOnItemClickListener(new
        AdapterView.OnItemClickListener() {
            @Override
            public void onItemClick(AdapterView<?> parent, View view, int
            position, long id) {
                User.chatWith = mArraylist.get(position);
                startActivity(new Intent(UsersListActivity.this,
                ChatActivity.class));
            }
        });

    }
```

When we make our Firebase endpoint URL public, anybody can read and write to the endpoint if they have the URL. I am just using the URL and adding `.json` as an extension so that it will return the JSON result. Now, we need to write one last method for managing the success result:

```
    public void doOnSuccess(String s){
        try {
            JSONObject obj = new JSONObject(s);

            Iterator i = obj.keys();
            String key = "";

            while(i.hasNext()){
                key = i.next().toString();

                if(!key.equals(User.username)) {
                    mArraylist.add(key);
                }

                totalUsers++;
            }

        } catch (JSONException e) {
            e.printStackTrace();
        }

        if(totalUsers <=1){
            mNoUsersText.setVisibility(View.VISIBLE);
            mUsersList.setVisibility(View.GONE);
        }
        else{
```

```
        mNoUsersText.setVisibility(View.GONE);
        mUsersList.setVisibility(View.VISIBLE);
        mUsersList.setAdapter(new ArrayAdapter<String>(this,
        android.R.layout.simple_list_item_1, mArraylist));
    }

    mProgressDialog.dismiss();
}
```

The complete class will look as follows:

```
public class UsersListActivity extends AppCompatActivity {

    private ListView mUsersList;
    private TextView mNoUsersText;
    private ArrayList<String> mArraylist = new ArrayList<>();
    private int totalUsers = 0;
    private ProgressDialog mProgressDialog;

    @Override
    protected void onCreate(Bundle savedInstanceState) {
        super.onCreate(savedInstanceState);
        setContentView(R.layout.activity_users_list);
        mUsersList = (ListView)findViewById(R.id.usersList);
        mNoUsersText = (TextView)findViewById(R.id.noUsersText);

        mProgressDialog = new ProgressDialog(UsersListActivity.this);
        mProgressDialog.setMessage("Loading...");
        mProgressDialog.show();

        String url = "https://packt-wear.firebaseio.com/users.json";
        StringRequest request = new StringRequest(Request.Method.GET,
        url, new Response.Listener<String>(){
            @Override
            public void onResponse(String s) {
                doOnSuccess(s);
            }
        },new Response.ErrorListener(){
            @Override
            public void onErrorResponse(VolleyError volleyError) {
                System.out.println("" + volleyError);
            }
        });

        RequestQueue rQueue =
        Volley.newRequestQueue(UsersListActivity.this);
        rQueue.add(request);
```

```
    mUsersList.setOnItemClickListener(new
    AdapterView.OnItemClickListener() {
        @Override
        public void onItemClick(AdapterView<?> parent, View view,
        int position, long id) {
            User.chatWith = mArraylist.get(position);
            startActivity(new Intent(UsersListActivity.this,
            ChatActivity.class));
        }
    });

}

public void doOnSuccess(String s){
    try {
        JSONObject obj = new JSONObject(s);

        Iterator i = obj.keys();
        String key = "";

        while(i.hasNext()){
            key = i.next().toString();

            if(!key.equals(User.username)) {
                mArraylist.add(key);
            }

            totalUsers++;
        }

    } catch (JSONException e) {
        e.printStackTrace();
    }

    if(totalUsers <=1){
        mNoUsersText.setVisibility(View.VISIBLE);
        mUsersList.setVisibility(View.GONE);
    }
    else{
        mNoUsersText.setVisibility(View.GONE);
        mUsersList.setVisibility(View.VISIBLE);
        mUsersList.setAdapter(new ArrayAdapter<String>(this,
        android.R.layout.simple_list_item_1, mArraylist));
    }

    mProgressDialog.dismiss();
    }
}
```

Now, we have completed one flow, which is onBoarding the user and showing a list of users who are available to chat. Now, it's time to work on the actual chatting logic. Let's start working on ChatActivity.

For the message background, we will be adding two drawable resources files: rounded_corner1.xml and rounded_corner2.xml. Let's add the XML code for the drawable resource files:

```
<?xml version="1.0" encoding="utf-8"?>
<shape xmlns:android="http://schemas.android.com/apk/res/android">
    <solid android:color="#dddddd" />
    <stroke
        android:width="0dip"
        android:color="#dddddd" />
    <corners android:radius="10dip" />
    <padding
        android:bottom="5dip"
        android:left="5dip"
        android:right="5dip"
        android:top="5dip" />
</shape>
```

For rounded_corner2.xml

```
<?xml version="1.0" encoding="utf-8"?>
<shape xmlns:android="http://schemas.android.com/apk/res/android">
    <solid android:color="#f0f0f0" />
    <stroke
        android:width="0dip"
        android:color="#f0f0f0" />
    <corners android:radius="10dip" />
    <padding
        android:bottom="5dip"
        android:left="5dip"
        android:right="5dip"
        android:top="5dip" />
</shape>
```

Now, let's declare the necessary instances for the Firebase chat activity:

```
//Global instances
LinearLayout mLinearlayout;
ImageView mSendButton;
EditText mMessageArea;
ScrollView mScrollview;
Firebase reference1, reference2;
```

```
//inside onCreate method
mLinearlayout = (LinearLayout)findViewById(R.id.layout1);
mSendButton = (ImageView)findViewById(R.id.sendButton);
mMessageArea = (EditText)findViewById(R.id.messageArea);
mScrollview = (ScrollView)findViewById(R.id.scrollView);

Firebase.setAndroidContext(this);
reference1 = new Firebase("https://packt-wear.firebaseio.com/messages/" +
User.username + "_" + User.chatWith);
reference2 = new Firebase("https://packt-wear.firebaseio.com/messages/" +
User.chatWith + "_" + User.username);
```

On clicking the **send** button, using the `push()` method, we can update Firebase with the username and with the message that they sent across:

```
mSendButton.setOnClickListener(new View.OnClickListener() {
    @Override
    public void onClick(View v) {
        String messageText = mMessageArea.getText().toString();

        if(!messageText.equals("")){
            Map<String, String> map = new HashMap<String, String>();
            map.put("message", messageText);
            map.put("user", User.username);
            reference1.push().setValue(map);
            reference2.push().setValue(map);
        }
    }
});
```

There are callbacks we need to implement from the Firebase `addChildEventListener()`. In the `onChildAdded` method, we can show the messages added. The following code completes Firebase and adds the background for the messages:

```
reference1.addChildEventListener(new ChildEventListener() {
    @Override
    public void onChildAdded(DataSnapshot dataSnapshot, String s) {
        Map map = dataSnapshot.getValue(Map.class);
        String message = map.get("message").toString();
        String userName = map.get("user").toString();

        if(userName.equals(User.username)){
            addMessageBox("You:-\n" + message, 1);
        }
        else{
            addMessageBox(User.chatWith + ":-\n" + message, 2);
        }
    }
```

```
@Override
public void onChildChanged(DataSnapshot dataSnapshot, String s) {

}

@Override
public void onChildRemoved(DataSnapshot dataSnapshot) {

}

@Override
public void onChildMoved(DataSnapshot dataSnapshot, String s) {

}

@Override
public void onCancelled(FirebaseError firebaseError) {

}
});
```

The `addMessageBox` method changes the sender and receiver message background:

```
public void addMessageBox(String message, int type){
TextView textView = new TextView(ChatActivity.this);
textView.setText(message);
LinearLayout.LayoutParams lp = new
LinearLayout.LayoutParams(ViewGroup.LayoutParams.MATCH_PARENT,
ViewGroup.LayoutParams.WRAP_CONTENT);
lp.setMargins(0, 0, 0, 10);
textView.setLayoutParams(lp);

    if(type == 1) {
textView.setBackgroundResource(R.drawable.rounded_corner1);
}
    else{
textView.setBackgroundResource(R.drawable.rounded_corner2);
}

    mLinearlayout.addView(textView);
    mScrollview.fullScroll(View.FOCUS_DOWN);
}
```

The complete code for `ChatActivity` is as follows:

```
public class ChatActivity extends AppCompatActivity {
```

```
LinearLayout mLinearlayout;
ImageView mSendButton;
EditText mMessageArea;
ScrollView mScrollview;
Firebase reference1, reference2;

@Override
protected void onCreate(Bundle savedInstanceState) {
    super.onCreate(savedInstanceState);
    setContentView(R.layout.activity_chat);

    mLinearlayout = (LinearLayout)findViewById(R.id.layout1);
    mSendButton = (ImageView)findViewById(R.id.sendButton);
    mMessageArea = (EditText)findViewById(R.id.messageArea);
    mScrollview = (ScrollView)findViewById(R.id.scrollView);

    Firebase.setAndroidContext(this);
    reference1 = new Firebase("https://packt-
    wear.firebaseio.com/messages/" + User.username + "_" +
    User.chatWith);
    reference2 = new Firebase("https://packt-
    wear.firebaseio.com/messages/" + User.chatWith + "_" +
    User.username);

    mSendButton.setOnClickListener(new View.OnClickListener() {
        @Override
        public void onClick(View v) {
            String messageText = mMessageArea.getText().toString();

            if(!messageText.equals("")){
                Map<String, String> map = new HashMap<String,
                String>();
                map.put("message", messageText);
                map.put("user", User.username);
                reference1.push().setValue(map);
                reference2.push().setValue(map);
            }
        }
    });

    reference1.addChildEventListener(new ChildEventListener() {
        @Override
        public void onChildAdded(DataSnapshot dataSnapshot,
        String s) {
            Map map = dataSnapshot.getValue(Map.class);
            String message = map.get("message").toString();
            String userName = map.get("user").toString();
```

```
            if(userName.equals(User.username)){
                addMessageBox("You:-\n" + message, 1);
            }
            else{
                addMessageBox(User.chatWith + ":-\n" + message, 2);
            }
        }

        @Override
        public void onChildChanged(DataSnapshot dataSnapshot,
        String s) {

        }

        @Override
        public void onChildRemoved(DataSnapshot dataSnapshot) {

        }

        @Override
        public void onChildMoved(DataSnapshot dataSnapshot,
        String s) {

        }

        @Override
        public void onCancelled(FirebaseError firebaseError) {

        }
    });
}

public void addMessageBox(String message, int type){
    TextView textView = new TextView(ChatActivity.this);
    textView.setText(message);
    LinearLayout.LayoutParams lp = new LinearLayout
    .LayoutParams(ViewGroup.LayoutParams.MATCH_PARENT,
    ViewGroup.LayoutParams.WRAP_CONTENT);
    lp.setMargins(0, 0, 0, 10);
    textView.setLayoutParams(lp);

    if(type == 1) {
        textView.setBackgroundResource(R.drawable.rounded_corner1);
    }
    else{
        textView.setBackgroundResource(R.drawable.rounded_corner2);
    }
```

```
        mLinearlayout.addView(textView);
        mScrollview.fullScroll(View.FOCUS_DOWN);
    }
}
```

We have the complete working chatting module for a mobile application. Now, let's write the Wear module for the chat application.

Wear App implementation

The objective of the wear module is that the wear device should receive new messages and show it in the app and users should be able to reply to that message from the wear device. In this, we will understand the classes and APIs for the Wear and mobile app to establish communications.

Now, the boilerplate code that Android Studio has generated a code for the timer app. All we need to do is delete all the code and just keep the onCreate() method. Later, in the activity_main.xml file, let's add the user interface that helps the user to chat. Here, I will have Edittext and Textview, and a button that sends the message to the mobile device. Let's add the XML code:

```xml
<?xml version="1.0" encoding="utf-8"?>
<android.support.wearable.view.BoxInsetLayout
xmlns:android="http://schemas.android.com/apk/res/android"
    xmlns:app="http://schemas.android.com/apk/res-auto"
    xmlns:tools="http://schemas.android.com/tools"
    android:id="@+id/container"
    android:layout_width="match_parent"
    android:layout_height="match_parent"
    tools:context="com.packt.smartchat.MainActivity"
    tools:deviceIds="wear">

    <LinearLayout
        android:layout_gravity="center|center_horizontal"
        android:layout_width="match_parent"
        android:layout_height="wrap_content"
        android:orientation="vertical">

        <TextView
            android:id="@+id/text"
            android:layout_width="match_parent"
            android:layout_height="match_parent"
            android:gravity="center"
            android:text="Message"
            android:layout_margin="30dp"
```

```
                app:layout_box="all" />

        <EditText
            android:id="@+id/message"
            android:layout_width="match_parent"
            android:layout_height="wrap_content" />


        <Button
            android:id="@+id/send"
            android:layout_gravity="center"
            android:text="Send"
            android:layout_width="wrap_content"
            android:layout_height="wrap_content" />
    </LinearLayout>


</android.support.wearable.view.BoxInsetLayout>
```

Now, the User Interface is ready. While we want to receive and send messages to mobiles, we need to write a service class that extends to `WearableListenerService` and the service class needs to be registered in the manifest:

```java
public class ListenerService extends WearableListenerService {

    @Override
    public void onMessageReceived(MessageEvent messageEvent) {

        if (messageEvent.getPath().equals("/message_path")) {
            final String message = new String(messageEvent.getData());
            Log.v("myTag", "Message path received on watch is: " +
            messageEvent.getPath());
            Log.v("myTag", "Message received on watch is: " + message);

            // Broadcast message to wearable activity for display
            Intent messageIntent = new Intent();
            messageIntent.setAction(Intent.ACTION_SEND);
            messageIntent.putExtra("message", message);
            LocalBroadcastManager
            .getInstance(this).sendBroadcast(messageIntent);
        }
        else {
            super.onMessageReceived(messageEvent);
        }
    }
}
```

Now, register the `service` class in the manifest file within the application tag scope as follows:

```
<application
    android:allowBackup="true"
    android:icon="@mipmap/ic_launcher"
    android:label="@string/app_name"
    android:supportsRtl="true"
    android:theme="@android:style/Theme.DeviceDefault">

<activity...></activity>

<service android:name=".ListenerService">
 <intent-filter>
 <action android:name="com.google.android.gms.wearable.DATA_CHANGED" />
 <action
 android:name="com.google.android.gms.wearable.MESSAGE_RECEIVED" />
 <data android:scheme="wear" android:host="*"
 android:pathPrefix="/message_path" />
 </intent-filter>
</service>

</application>
```

We are registering the `wear` service with the newest standard. Earlier, we had to register the service using the `BIND_LISTENER` API. Due to its inefficiency, it is deprecated. We have to use the previous `DATA_CHANGED` and `MESSAGE_RECEIVED` APIs, since they let the app listen to a particular path. Whereas the `BIND_LISTENER` API was listening to the wide range of system messages and that was a performance and battery level drawback. The following code illustrates the deprecated `BIND_LISTENER` registration:

```
<service android:name=".ListenerService">
<intent-filter>
<action android:name="com.google.android.gms.wearable.BIND_LISTENER" />
</intent-filter>
</service>
```

After registering the service in the manifest, we can directly work `MainActivity` in the Wear module. Before we get started, make sure you have removed all the boilerplate code in `MainActivity` with only an `onCreate` method, which looks as follows:

```
public class MainActivity extends WearableActivity {

    @Override
    protected void onCreate(Bundle savedInstanceState) {
        super.onCreate(savedInstanceState);
        setContentView(R.layout.activity_main);
```

```
    setAmbientEnabled();

  }
}
```

Connect all the XML components in `MainActivity`:

```
mTextView = (TextView) findViewById(R.id.text);
send = (Button) findViewById(R.id.send);
message = (EditText) findViewById(R.id.message);
```

Let's implement the interfaces from `GoogleApiClient` that helps in finding connected nodes and for handling failure scenarios:

```
public class MainActivity extends WearableActivity implements
GoogleApiClient.ConnectionCallbacks,
        GoogleApiClient.OnConnectionFailedListener  {


        . . . . . . .


}
```

After implementing `ConnectionCallbacks` and `OnConnectionFailedListener`, we have to override a few methods from this interface, namely, the `onConnected`, `onConnectionSuspended`, and `onConnectionFailed` methods. Most of our logic will be programmed in the `onConnected` method. Now, inside the `MainActivity` scope, we need to write a class that extends `BroadcastReciever` with the overriding `onReceive` method for listening to the message:

```
public class MessageReceiver extends BroadcastReceiver {
    @Override
    public void onReceive(Context context, Intent intent) {
        String message = intent.getStringExtra("message");
        Log.v("packtchat", "Main activity received message: " +
        message);
        // Display message in UI
        mTextView.setText(message);
    }
}
```

Register the Local Broadcast receiver in the `onCreate` method as follows:

```
// Register the local broadcast receiver
IntentFilter messageFilter = new IntentFilter(Intent.ACTION_SEND);
MessageReceiver messageReceiver = new MessageReceiver();
LocalBroadcastManager.getInstance(this).registerReceiver(messageReceiver,
messageFilter);
```

Declare `GoogleApiClient` and Node instances with `WEAR_PATH` for sending messages from wear that the mobile app will listen to:

```
private Node mNode;
private GoogleApiClient mGoogleApiClient;
private static final String WEAR_PATH = "/from-wear";
```

And initialize the `mGoogleApiclient` in the `onCreate` method:

```
//Initialize mGoogleApiClient
mGoogleApiClient = new GoogleApiClient.Builder(this)
        .addApi(Wearable.API)
        .addConnectionCallbacks(this)
        .addOnConnectionFailedListener(this)
        .build();
```

In the `onConnected` method, we will use the wearable node API for fetching all the connected nodes. It can be a wear or mobile device, which is paired. Using the following code, we will know which wear devices are paired:

```
@Override
public void onConnected(@Nullable Bundle bundle) {
    Wearable.NodeApi.getConnectedNodes(mGoogleApiClient)
            .setResultCallback(new
            ResultCallback<NodeApi.GetConnectedNodesResult>() {
                @Override
                public void onResult(NodeApi.GetConnectedNodes
                Result nodes) {
                    for (Node node : nodes.getNodes()) {
                        if (node != null && node.isNearby()) {
                            mNode = node;
                            Log.d("packtchat", "Connected to " +
                            mNode.getDisplayName());
                        }
                    }
                    if (mNode == null) {
                        Log.d("packtchat", "Not connected!");
                    }
                }
            });
}
```

Now, we need to attach `clicklistener` to the `send` button instance for fetching the value from `edittext` and passing it to a method that sends the message to the mobile device:

```
send.setOnClickListener(new View.OnClickListener() {
    @Override
    public void onClick(View v) {
        mMsgStr = message.getText().toString();
        sendMessage(mMsgStr);
    }
});
```

The `sendMessage` method takes one string argument and sends the same string message to the connected node as bytes. Using `MessageAPI`, we will send the message to the mobile. The following code explains how this is achieved:

```
private void sendMessage(String city) {
    if (mNode != null && mGoogleApiClient != null) {
        Wearable.MessageApi.sendMessage(mGoogleApiClient,
        mNode.getId(), WEAR_PATH, city.getBytes())
                .setResultCallback(new
                ResultCallback<MessageApi.SendMessageResult>() {
                    @Override
                    public void onResult(MessageApi.SendMessageResult
                    sendMessageResult) {
                        if (!sendMessageResult.getStatus().isSuccess())
                        {
                            Log.d("packtchat", "Failed message: " +
                            sendMessageResult.getStatus()
                            .getStatusCode());
                        } else {
                            Log.d("packtchat", "Message succeeded");
                        }
                    }
                });
    }
}
```

Let's override the `onstart` and `onstop` methods for seeking help in connecting and disconnecting `GoogleAPIClient`:

```
@Override
protected void onStart() {
    super.onStart();
    mGoogleApiClient.connect();
}

@Override
```

```
protected void onStop() {
    super.onStop();
    mGoogleApiClient.disconnect();
}
```

The complete wear module `MainActivity` code is as follows:

```
public class MainActivity extends WearableActivity implements
GoogleApiClient.ConnectionCallbacks,
        GoogleApiClient.OnConnectionFailedListener  {

    private TextView mTextView;
    Button send;
    EditText message;
    String mMsgStr;

    private Node mNode;
    private GoogleApiClient mGoogleApiClient;
    private static final String WEAR_PATH = "/from-wear";


    @Override
    protected void onCreate(Bundle savedInstanceState) {
        super.onCreate(savedInstanceState);
        setContentView(R.layout.activity_main);

        mTextView = (TextView)findViewById(R.id.text);
        send = (Button) findViewById(R.id.send);
        message = (EditText)findViewById(R.id.message);
        setAmbientEnabled();

        //Initialize mGoogleApiClient
        mGoogleApiClient = new GoogleApiClient.Builder(this)
                .addApi(Wearable.API)
                .addConnectionCallbacks(this)
                .addOnConnectionFailedListener(this)
                .build();

        // Register the local broadcast receiver
        IntentFilter messageFilter = new
        IntentFilter(Intent.ACTION_SEND);
        MessageReceiver messageReceiver = new MessageReceiver();
        LocalBroadcastManager.getInstance(this)
        .registerReceiver(messageReceiver, messageFilter);

        send.setOnClickListener(new View.OnClickListener() {
            @Override
            public void onClick(View v) {
```

```
                mMsgStr = message.getText().toString();
                sendMessage(mMsgStr);
            }
        });

    }

    private void sendMessage(String message) {
        if (mNode != null && mGoogleApiClient != null) {
            Wearable.MessageApi.sendMessage(mGoogleApiClient,
            mNode.getId(), WEAR_PATH, message.getBytes())
                    .setResultCallback(new
                    ResultCallback<MessageApi.SendMessageResult>() {
                        @Override
                        public void onResult(MessageApi
                        .SendMessageResult sendMessageResult) {
                            if (!sendMessageResult.getStatus()
                            .isSuccess()) {
                                Log.d("packtchat", "Failed message: " +
                                sendMessageResult.getStatus()
                                .getStatusCode());
                            } else {
                                Log.d("packtchat", "Message
                                succeeded");
                            }
                        }
                    });
        }
    }


    public class MessageReceiver extends BroadcastReceiver {
        @Override
        public void onReceive(Context context, Intent intent) {
            String message = intent.getStringExtra("message");
            Log.v("packtchat", "Main activity received message: " +
            message);
            // Display message in UI
            mTextView.setText(message);

        }
    }

    @Override
    public void onConnected(@Nullable Bundle bundle) {
        Wearable.NodeApi.getConnectedNodes(mGoogleApiClient)
                .setResultCallback(new
                ResultCallback<NodeApi.GetConnectedNodesResult>() {
```

```
                        @Override
                        public void
                        onResult(NodeApi.GetConnectedNodesResult nodes) {
                            for (Node node : nodes.getNodes()) {
                                if (node != null && node.isNearby()) {
                                    mNode = node;
                                    Log.d("packtchat", "Connected to " +
                                    mNode.getDisplayName());
                                }
                            }
                            if (mNode == null) {
                                Log.d("packtchat", "Not connected!");
                            }
                        }
                    });
        }

        @Override
        protected void onStart() {
            super.onStart();
            mGoogleApiClient.connect();
        }

        @Override
        protected void onStop() {
            super.onStop();
            mGoogleApiClient.disconnect();
        }

        @Override
        public void onConnectionSuspended(int i) {

        }

        @Override
        public void onConnectionFailed(@NonNull ConnectionResult
        connectionResult) {

        }
    }
```

The Wear module is ready to receive and send the message to the connected device. Now, we need to upgrade our mobile module to receive and send the messages to the wear.

Switch to the mobile module and create a service class `WearListner` and override the `onMessageReceived` method as follows:

```
public class WearListner extends WearableListenerService {

    @Override
    public void onMessageReceived(MessageEvent messageEvent) {
        if (messageEvent.getPath().equals("/from-wear")) {
            final String message = new String(messageEvent.getData());
            Log.v("pactchat", "Message path received on watch is: " +
            messageEvent.getPath());
            Log.v("packtchat", "Message received on watch is: " +
            message);

            // Broadcast message to wearable activity for display
            Intent messageIntent = new Intent();
            messageIntent.setAction(Intent.ACTION_SEND);
            messageIntent.putExtra("message", message);
            LocalBroadcastManager
            .getInstance(this).sendBroadcast(messageIntent);
        }
        else {
            super.onMessageReceived(messageEvent);
        }
    }
}
```

Now, register this `WearListner` class in the manifest within the application tag scope:

```
<service android:name=".WearListner">
    <intent-filter>
        <action
        android:name="com.google.android.gms.wearable.DATA_CHANGED" />
        <action
        android:name="com.google.android.gms.wearable.MESSAGE_RECEIVED"
        />
        <data android:scheme="wear" android:host="*"
android:pathPrefix="/from-wear" />
    </intent-filter>
</service>
```

Now let's switch our work-scope to mobile module and add the following changes for deeplinking with wear and mobile.
Implement the `ConnectionCallbacks` and `OnConnectionFailedListener` interfaces from the `GoogleApiClient` class in `ChatActivity`:

```
public class ChatActivity extends AppCompatActivity implements
```

```
            GoogleApiClient.ConnectionCallbacks,
            GoogleApiClient.OnConnectionFailedListener {
    ...
    }
```

Override the methods from the `ConnectionCallbacks` and `OnConnectionFailedListener` interfaces similar to what we did for wear `MainActivity`:

Initialize `GoogleApiClient` in the `onCreate` method as follows:

```
    googleClient = new GoogleApiClient.Builder(this)
            .addApi(Wearable.API)
            .addConnectionCallbacks(this)
            .addOnConnectionFailedListener(this)
            .build();
```

Write the broadcast receiver class within `ChatActivity` along with the string message we received. We need to pass it in the `addMessageBox` method, which we have written already:

```
    public class MessageReceiver extends BroadcastReceiver {
        @Override
        public void onReceive(Context context, Intent intent) {
            String message = intent.getStringExtra("message");
            Log.v("myTag", "Main activity received message: " + message);
            // Displaysage in UI
            addMessageBox("You:-\n" + message, 1);
            if(!message.equals("")){
                Map<String, String> map = new HashMap<String, String>();
                map.put("message", message);
                map.put("user", User.username);
                reference1.push().setValue(map);
            }
            new SendToDataLayerThread("/message_path","You:-\n" +
            message).start();
        }
    }
```

Register `MessageReciever` in the `onCreate` method as follows:

```
    // Register the local broadcast receiver
    IntentFilter messageFilter = new IntentFilter(Intent.ACTION_SEND);
    MessageReceiver messageReceiver = new MessageReceiver();
    LocalBroadcastManager.getInstance(this).registerReceiver(messageReceiver,
    messageFilter);
```

After registering the broadcast receiver, write the `SendToDataLayerThread` class that extends to the `Thread` class for taking all the load in the separate thread, but on UI Thread. In the `void run` method, we will check for all the connected nodes and loop through the connected node. Once the connection is established, we will use `MessageAPI` to send the message as shown in the code. The Message API `sendMessage` method looks for certain parameters, such as `googleclient`, and the connected Node ID path, exactly what we registered in the wear manifest and the actual message as bytes. Using the `SendMessageResult` instance, we developers can make sure of whether the message fired from the device reached the nodes successfully or not:

```
class SendToDataLayerThread extends Thread {
    String path;
    String message;

    // Constructor to send a message to the data layer
    SendToDataLayerThread(String p, String msg) {
        path = p;
        message = msg;
    }

    public void run() {
        NodeApi.GetConnectedNodesResult nodes =
        Wearable.NodeApi.getConnectedNodes(googleClient).await();
        for (Node node : nodes.getNodes()) {
            MessageApi.SendMessageResult result = Wearable.MessageApi
            .sendMessage(googleClient, node.getId(), path,
            message.getBytes()).await();
            if (result.getStatus().isSuccess()) {
                Log.v("myTag", "Message: {" + message + "} sent to: " +
                node.getDisplayName());
            } else {
                // Log an error
                Log.v("myTag", "ERROR: failed to send Message");
            }
        }
    }
}
```

We need to initialize the `sendtoDatalayer` thread in a the few of methods in `chatActivity`:

```
@Override
public void onConnected(@Nullable Bundle bundle) {
    String message = "Conected";
    //Requires a new thread to avoid blocking the UI
    new SendToDataLayerThread("/message_path", message).start();
```

```
    }
```

When `reference1` is updated with some child events, we need to add the message to the `sendToDatalayer` thread:

```
reference1.addChildEventListener(new ChildEventListener() {
    @Override
    public void onChildAdded(DataSnapshot dataSnapshot, String s) {
        Map map = dataSnapshot.getValue(Map.class);
        String message = map.get("message").toString();
        String userName = map.get("user").toString();
        mMessageArea.setText("");
        if(userName.equals(User.username)){
            addMessageBox("You:-\n" + message, 1);
            new SendToDataLayerThread("/message_path","You:-\n" +
            message).start();

        }
        else{
            addMessageBox(User.chatWith + ":-\n" + message, 2);
            new SendToDataLayerThread("/message_path", User.chatWith +
            ":-\n" + message).start();

        }
    }
```

Add the following callbacks for connecting and disconnecting the `GoogleApiClient`, Add the callbacks in the `ChatActivity`:

```
@Override
protected void onStart() {
    super.onStart();
    googleClient.connect();
}

@Override
protected void onStop() {
    if (null != googleClient && googleClient.isConnected()) {
        googleClient.disconnect();
    }
    super.onStop();
}
```

The chatting application is complete with wear and mobile interaction. Every message that chat activity receives is sent to wear and the reply from wear is updated to the mobile device.

The chat screen that shows the basic conversation between two users looks as follows:

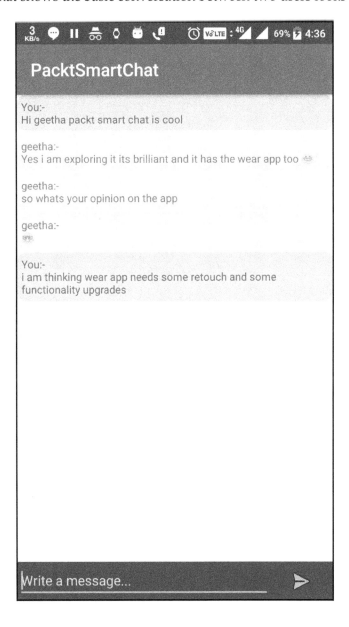

The wear app screen will look as follows in a round dial watch:

Summary

In this chapter, we have seen how we can utilize the Firebase real-time database as a chatting medium. We assembled a straightforward messaging application that can send a reply from a wear device. There is ample space to this project to enhance the elements of the project. We have seen how a chatting application can send and receive messages crosswise over wear and handheld devices. We have conceptualised a chatting application from scratch and we have set Data Layer events for the nodes to communicate with each other. The basic idea of Messaging API is to strengthen our understanding of wearable communication. Also the `GoogleApiClient` class plays a noteworthy part in Play services.

In the next chapter, we will understand notifications, Firebase functions, and how push notifications can be triggered using Firebase functions.

9
Let us Chat in a Smart Way - Notifications and More

The previous chapter helped us in building a conversational messaging application, but the Wear app had a very normal user interface with no notifications whatsoever. Notification is a very important aspect of a messaging application, but it needs complex infrastructure to process the notification. When the sender sends a message to the receiver, the receiver should get a notification conveying certain information, such as the sender name, and a quick message preview.

Notification is a component in Android that we can use to display information. In the case of a messaging application, the receiver should get the push notification to instantiate the notification component. So, whenever there is a real-time database update in Firebase, both the handheld device and the Wear device should get the notification. Thankfully, we don't need a server to handle the notification; Firebase will handle push notifications for your messaging application. When there is an update in the real-time database, we need to fire a push notification.

In this chapter, we will explore the following:

- Firebase functions
- Notifications
- Material design Wear app

Firebase functions

Firebase functions are the smartest solution for monitoring database triggers and many more server-related executions. Instead of hosting a server, which listens to Firebase database changes and then fires the push notification, we shall utilize one of the Firebase technologies to complete this task. Firebase functions has an efficient control on all the Firebase technologies, such as Firebase authentication, storage, analytics, and so on. Firebase functions can be used in various aspects; for example, when your analytic cohorts meet some milestone, you can send targeted notifications, invites, and so on. Any server level business logic that you might want to implement in your Firebase system has been made with Firebase functions. We will use Firebase functions for sending push notifications whenever a database trigger occurs. We need an entry level JavaScript understanding to complete this task.

To get started on Firebase functions, we need Node.js environment that you can install it from `https://nodejs.org/`. When you install Node.js, it will also install **node package manager** (**npm**). It helps install the JavaScript frameworks, plugins, and so on. After installing Node.js, go to the terminal or your command line.

Check whether node is installed by entering `$node --version`. If the CLI returns the latest version number, then you are good to go:

```
//Install Firebase CLI
$ npm install -g firebase-tools
```

If you encounter any errors, you should execute the command in the super user mode:

```
sudo su
npm install -g firebase-tools
```

Navigate to a directory in which you wish to save the Firebase function programs and authenticate yourself using following command:

```
//CLI authentication
$ firebase login
```

After successful authentication you can initialize the Firebase functions:

```
//Initialise Firebase functions
$ firebase init functions
```

The CLI tool will generate the Firebase function and the necessary code in the directory you initialized. Open `index.js` in your favorite editor. We create a Firebase function for the `Realtime` database events with `functions.database` to handle the real-time database triggers. We shall call ref(path) to reach to the particular database `path`. `onwrite()` method from the functions, which will send notifications whenever there is an update in the database.

Now, to construct our notification payload, let's understand our real-time database structure:

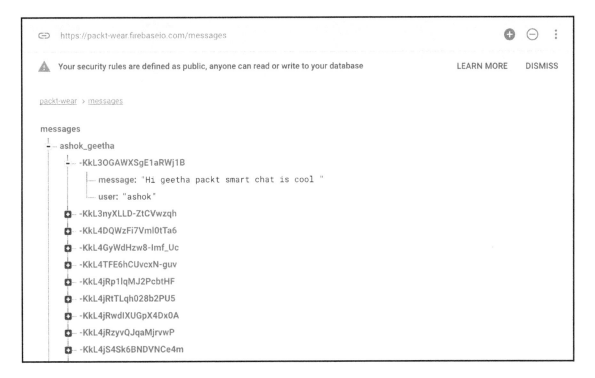

We can see that **messages** have a child named `ashok_geetha`, which conveys that these two users' conversations are stored inside with a unique Firebase ID. For this implementation, we will pick `ashok_geetha` for push notification.

Now, in the `index.js` file, add the following code:

```
//import firebase functions modules
const functions = require('firebase-functions');
//import admin module
        const admin = require('firebase-admin');
        admin.initializeApp(functions.config().firebase);
```

```
// Listens for new messages added to messages/:pushId
exports.pushNotification = functions.database
.ref('/messages/ashok_geetha/{pushId}').onWrite( event => {

console.log('Push notification event triggered');

//  Grab the current value of what was written to the Realtime
Database.
var valueObject = event.data.val();

if(valueObject.photoUrl != null) {
valueObject.photoUrl= "Sent you a lot of love!";
}

// Create a notification
const payload = {
notification: {
title:valueObject.message,
body: valueObject.user || valueObject.photoUrl,
sound: "default"
},
};

//Create an options object that contains the time to live for
the notification and the priority
const options = {
priority: "high",
timeToLive: 60 * 60 * 24
};


return admin.messaging().sendToTopic("pushNotifications",
payload, options);
});
```

When we have different firebase structure, using the url-id configuration
we can trigger the firebase functions for making the pushnotification to
work with all your users. We just need to make the following change in
the url. /messages/{ChatID}/{pushId}

Now, in the terminal, complete the deployment of Firebase functions with the $ firebase deploy command:

```
sh-3.2# firebase deploy

=== Deploying to 'packt-wear'...

i  deploying functions
i  functions: ensuring necessary APIs are enabled...
i  runtimeconfig: ensuring necessary APIs are enabled...
✓  runtimeconfig: all necessary APIs are enabled
✓  functions: all necessary APIs are enabled
i  functions: preparing functions directory for uploading...
i  functions: packaged functions (1.2 KB) for uploading
✓  functions: functions folder uploaded successfully
i  starting release process (may take several minutes)...
i  functions: updating function pushNotification...
✓  functions[pushNotification]: Successful update operation.
✓  functions: all functions deployed successfully!

✓  Deploy complete!

Project Console: https://console.firebase.google.com/project/packt-wear/overview
```

The previous Node.js setup fires the push notification using the Firebase functions. Now, we need a mechanism to receive the message from the Firebase in our local mobile and Wear application.

Switch to Android Studio and add the following dependency to your mobile project gradle module file:

```
//Add the dependency
dependencies {
    compile 'com.google.firebase:firebase-messaging:10.2.4'
}
```

After adding the dependency, create a class called MessagingService, which extends to the FirebaseMessagingService class. The FirebaseMessaging class extends to com.google.firebase.iid.zzb and the zzb class extends to Android Service class. This class will help the communication process between Firebase messaging and Android application. It also can display notifications automatically. Let's create the class and extend it to the FirebaseMessagingService class, as follows:

```
public class MessagingService extends FirebaseMessagingService {

    //Override methods
}
```

It's time to add the override method `onMessageReceived`. This method receives the notification when app is at the foreground or background and we can retrieve all the notification parameters with the `getnotification()` method:

```
@Override
public void onMessageReceived(RemoteMessage remoteMessage) {
    super.onMessageReceived(remoteMessage);
}
```

The `RemoteMessage` object will have all the data we sent in the notification payload from Firebase functions. Inside the method, add the following code to get the title and message content. We are sending a message in the title param on the Firebase functions; you can customize it as per your use cases:

```
String notificationTitle = null, notificationBody = null;
// Check if message contains a notification payload.
if (remoteMessage.getNotification() != null) {
    Log.d(TAG, "Message Notification Body: " +
remoteMessage.getNotification().getBody());
    notificationTitle = remoteMessage.getNotification().getTitle();
    notificationBody = remoteMessage.getNotification().getBody();
}
```

For constructing notification, we will use `NotificationCompat.Builder` and, when a user clicks on the notification, we will take him to `MainActivity`:

```
private void sendNotification(String notificationTitle, String
notificationBody) {
    Intent intent = new Intent(this, MainActivity.class);
    intent.addFlags(Intent.FLAG_ACTIVITY_CLEAR_TOP);
    PendingIntent pendingIntent = PendingIntent.getActivity(this, 0,
    intent,
            PendingIntent.FLAG_ONE_SHOT);

    Uri defaultSoundUri= RingtoneManager.getDefaultUri
    (RingtoneManager.TYPE_NOTIFICATION);
    NotificationCompat.Builder notificationBuilder =
    (NotificationCompat.Builder) new NotificationCompat.Builder(this)
            .setAutoCancel(true)    //Automatically delete the
                                    notification
            .setSmallIcon(R.mipmap.ic_launcher) //Notification icon
            .setContentIntent(pendingIntent)
            .setContentTitle(notificationTitle)
            .setContentText(notificationBody)
            .setSound(defaultSoundUri);
```

```
    NotificationManager notificationManager = (NotificationManager)
    getSystemService(Context.NOTIFICATION_SERVICE);

    notificationManager.notify(0, notificationBuilder.build());
}
```

Call the method inside `onMessageReceived` and pass the content to the `sendNotification` method:

```
sendNotification(notificationTitle, notificationBody);
```

The complete class code will look as follows:

```
public class MessagingService extends FirebaseMessagingService {

    private static final String TAG = "MessagingService";

    @Override
    public void onMessageReceived(RemoteMessage remoteMessage) {
        String notificationTitle = null, notificationBody = null;
        // Check if message contains a notification payload.
        if (remoteMessage.getNotification() != null) {
            Log.d(TAG, "Message Notification Body: " +
            remoteMessage.getNotification().getBody());
            notificationTitle =
            remoteMessage.getNotification().getTitle();
            notificationBody =
            remoteMessage.getNotification().getBody();

            sendNotification(notificationTitle, notificationBody);

        }
    }

    private void sendNotification(String notificationTitle, String
    notificationBody) {
        Intent intent = new Intent(this, MainActivity.class);
        intent.addFlags(Intent.FLAG_ACTIVITY_CLEAR_TOP);
        PendingIntent pendingIntent = PendingIntent.getActivity(this,
        0, intent,
                PendingIntent.FLAG_ONE_SHOT);

        Uri defaultSoundUri= RingtoneManager.getDefaultUri
        (RingtoneManager.TYPE_NOTIFICATION);
        NotificationCompat.Builder notificationBuilder =
        (NotificationCompat.Builder)
        new NotificationCompat.Builder(this)
                .setAutoCancel(true)    //Automatically delete the
```

```
                              notification
            .setSmallIcon(R.mipmap.ic_launcher) //Notification icon
            .setContentIntent(pendingIntent)
            .setContentTitle(notificationTitle)
            .setContentText(notificationBody)
            .setSound(defaultSoundUri);


        NotificationManager notificationManager = (NotificationManager)
    getSystemService(Context.NOTIFICATION_SERVICE);

        notificationManager.notify(0, notificationBuilder.build());
    }
  }
```

We can now register the previous service in the manifest:

```
<service
    android:name=".MessagingService">
    <intent-filter>
        <action android:name="com.google.firebase.MESSAGING_EVENT"/>
    </intent-filter>
</service>
```

After all, we now have a service for listening to `pushNotification` but we need to listen
to the particular string we are sending. We can add the string to some constants or to the
XML file, but when we ask Firebase to send particular channel notifications, then we need
to subscribe to a channel that is called topic. Add the following code in `ChatActivity` and
inside the method:

```
FirebaseMessaging.getInstance().subscribeToTopic("pushNotifications");
```

To make the earlier operation global, create a class that extends to the `Application` class.
Inside the `oncreate` method, we can subscribe to the topic as follows:

```
public class PacktApp extends Application {

    @Override
    public void onCreate() {
        super.onCreate();
        Firebase.setAndroidContext(this);
        FirebaseMessaging.getInstance()
        .subscribeToTopic("pushNotifications");
    }
}
```

Now, we need to register the application class in the manifest. The application class holds the control on the onCreate method of the application lifecycle and it will help in maintaining the application life state:

```
<application
    android:name=".PacktApp"
    ...>

</application>
```

Congratulations! We have successfully configured push notification and are receiving it on our mobile phones:

In the connected Wear device, we will be able to see the following notification when we receive it. By default, the `NotificationCompat.Builder` class will help Wear devices to receive notifications, and if we want to customize it, we can do so by following the coming section.

From the Wear notification component, we can receive the reply right from the Wear device to the mobile application. To be able to achieve this, we will use the `WearExtender` component from the `NotificationCompat` class. Using this setup, users will be able to access the voice input, **Input method framework (IMF)**, and Emoji to reply to the notification:

```
Notification notification = new NotificationCompat.Builder(mContext)
        .setContentTitle("New mail from " + sender.toString())
        .setContentText(subject)
        .setSmallIcon(R.drawable.new_mail)
        .extend(new NotificationCompat.WearableExtender()
                .setContentIcon(R.drawable.new_mail))
        .build();
NotificationManagerCompat.from(mContext).notify(0, notification);
```

There will be numerous use cases where we need to send a quick response with already stored replies and or quick typing facility. In that case, we can make use of `Action.WearableExtender`:

```
/Android Wear requires a hint to display the reply action inline.
Action.WearableExtender actionExtender =
    new Action.WearableExtender()
        .setHintLaunchesActivity(true)
        .setHintDisplayActionInline(true);
wearableExtender.addAction(actionBuilder.extend(actionExtender).build());
```

Now, in the project, let's update our messaging service class, where we are firing the notification component when the background service receives a push notification.

Add the following instances to global scope of the class:

```
private static final String TAG = "MessagingService";
public static final String EXTRA_VOICE_REPLY = "extra_voice_reply";
public static final String REPLY_ACTION =
        "com.packt.smartcha.ACTION_MESSAGE_REPLY";
```

When a reply is received from the Wear, we will keep that in reference and pass it to the notification handler:

```
// Creates an Intent that will be triggered when a reply is received.
private Intent getMessageReplyIntent(int conversationId) {
    return new Intent().setAction(REPLY_ACTION).putExtra("1223",
    conversationId);
}
```

Wear notification is evolving with every new Android release and, right within the `NotificationCompat.Builder`, we have all the features that can make your mobile app interact with the Wear device. When you have a mobile app and it has interactions, such as notifications and so on, from the Wear device, you can get textual replies even if you don't have a Wear companion application.

Messaging service class

In the `MessagingService` class, we have a method called `sendNotification` that fires a notification to Wear and mobile devices. Let's update the method with the following code changes:

```
private void sendNotification(String notificationTitle, String
notificationBody) {
    // Wear 2.0 allows for in-line actions, which will be used for
    "reply".
    NotificationCompat.Action.WearableExtender inlineActionForWear2 =
            new NotificationCompat.Action.WearableExtender()
                    .setHintDisplayActionInline(true)
                    .setHintLaunchesActivity(false);

    RemoteInput remoteInput = new
    RemoteInput.Builder("extra_voice_reply").build();


    // Building a Pending Intent for the reply action to trigger.
    PendingIntent replyIntent = PendingIntent.getBroadcast(
            getApplicationContext(),
            0,
            getMessageReplyIntent(1),
            PendingIntent.FLAG_UPDATE_CURRENT);

    // Add an action to allow replies.
    NotificationCompat.Action replyAction =
            new NotificationCompat.Action.Builder(
                    R.drawable.googleg_standard_color_18,
```

```
                             "Notification",
                             replyIntent)

                             /// TODO: Add better wear support.
                             .addRemoteInput(remoteInput)
                             .extend(inlineActionForWear2)
                             .build();

            Intent intent = new Intent(this, ChatActivity.class);
            intent.addFlags(Intent.FLAG_ACTIVITY_CLEAR_TOP);
            PendingIntent pendingIntent = PendingIntent.getActivity(this, 0,
            intent,
                    PendingIntent.FLAG_ONE_SHOT);

            Uri defaultSoundUri = RingtoneManager.getDefaultUri
            (RingtoneManager.TYPE_NOTIFICATION);
            NotificationCompat.Builder notificationBuilder =
            (NotificationCompat.Builder)
            new NotificationCompat.Builder(this)
                    .setAutoCancel(true)    //Automatically delete the
                    notification
                    .setSmallIcon(R.mipmap.ic_launcher) //Notification icon
                    .setContentIntent(pendingIntent)
                    .addAction(replyAction)
                    .setContentTitle(notificationTitle)
                    .setContentText(notificationBody)
                    .setSound(defaultSoundUri);


            NotificationManagerCompat notificationManager =
            NotificationManagerCompat.from(this);
            notificationManager.notify(0, notificationBuilder.build());
    }
```

The previous methodwhich features in Wear device to have the IMF input, voice reply, and drawing emoji symbols. The complete class after modifying the code looks as follows:

```
package com.packt.smartchat;

/**
 * Created by ashok.kumar on 02/06/17.
 */

public class MessagingService extends FirebaseMessagingService {

    private static final String TAG = "MessagingService";
    public static final String EXTRA_VOICE_REPLY = "extra_voice_reply";
    public static final String REPLY_ACTION =
```

```
        "com.packt.smartcha.ACTION_MESSAGE_REPLY";
        public static final String SEND_MESSAGE_ACTION =
        "com.packt.smartchat.ACTION_SEND_MESSAGE";

@Override
public void onMessageReceived(RemoteMessage remoteMessage) {
    String notificationTitle = null, notificationBody = null;
    // Check if message contains a notification payload.
    if (remoteMessage.getNotification() != null) {
        Log.d(TAG, "Message Notification Body: " +
        remoteMessage.getNotification().getBody());
        notificationTitle =
        remoteMessage.getNotification().getTitle();
        notificationBody =
        remoteMessage.getNotification().getBody();

        sendNotification(notificationTitle, notificationBody);

    }
}

// Creates an intent that will be triggered when a message is read.
private Intent getMessageReadIntent(int id) {
    return new Intent().setAction("1").putExtra("1482", id);
}

// Creates an Intent that will be triggered when a reply is
received.
private Intent getMessageReplyIntent(int conversationId) {
    return new Intent().setAction(REPLY_ACTION).putExtra("1223",
    conversationId);
}

private void sendNotification(String notificationTitle, String
notificationBody) {
    // Wear 2.0 allows for in-line actions, which will be used for
    "reply".
    NotificationCompat.Action.WearableExtender
    inlineActionForWear2 =
            new NotificationCompat.Action.WearableExtender()
                    .setHintDisplayActionInline(true)
                    .setHintLaunchesActivity(false);

    RemoteInput remoteInput = new
    RemoteInput.Builder("extra_voice_reply").build();


    // Building a Pending Intent for the reply action to trigger.
```

```
    PendingIntent replyIntent = PendingIntent.getBroadcast(
            getApplicationContext(),
            0,
            getMessageReplyIntent(1),
            PendingIntent.FLAG_UPDATE_CURRENT);

    // Add an action to allow replies.
    NotificationCompat.Action replyAction =
            new NotificationCompat.Action.Builder(
                    R.drawable.googleg_standard_color_18,
                    "Notification",
                    replyIntent)

                    /// TODO: Add better wear support.
                    .addRemoteInput(remoteInput)
                    .extend(inlineActionForWear2)
                    .build();

    Intent intent = new Intent(this, ChatActivity.class);
    intent.addFlags(Intent.FLAG_ACTIVITY_CLEAR_TOP);
    PendingIntent pendingIntent = PendingIntent.getActivity(this,
    0, intent,
            PendingIntent.FLAG_ONE_SHOT);

    Uri defaultSoundUri = RingtoneManager.getDefaultUri
      (RingtoneManager.TYPE_NOTIFICATION);
    NotificationCompat.Builder notificationBuilder =
    (NotificationCompat.Builder) new
     NotificationCompat.Builder(this)
            .setAutoCancel(true)    //Automatically delete the
            notification
            .setSmallIcon(R.mipmap.ic_launcher) //Notification icon
            .setContentIntent(pendingIntent)
            .addAction(replyAction)
            .setContentTitle(notificationTitle)
            .setContentText(notificationBody)
            .setSound(defaultSoundUri);


    NotificationManagerCompat notificationManager =
    NotificationManagerCompat.from(this);
    notificationManager.notify(0, notificationBuilder.build());
    }
}
```

The notification received will look as follows:

When the user clicks on the notification, he or she will be given three options, as follows:

After receiving the notification on the Wear device, the user will reply with his or her thoughts via text or voice input or with the help of emojis. To handle this scenario, we need to write one broadcast receiver. Let's create a class called `MessageReplyReceiver`, extend it to the `BroadcastReceiver` class, and override the `onReceive` method. When you get the reply, just update the intent with the `conversationId`. `onReceive` method as follows:

```
@Override
public void onReceive(Context context, Intent intent) {
    if (MessagingService.REPLY_ACTION.equals(intent.getAction())) {
        int conversationId = intent.getIntExtra("reply", -1);
        CharSequence reply = getMessageText(intent);
```

```
        if (conversationId != -1) {
            Log.d(TAG, "Got reply (" + reply + ") for ConversationId "
            + conversationId);
        }
        // Tell the Service to send another message.
        Intent serviceIntent = new Intent(context,
        MessagingService.class);
        serviceIntent.setAction(MessagingService.SEND_MESSAGE_ACTION);
        context.startService(serviceIntent);
    }
}
```

From the `remoteIntent` object, to receive the data and convert the intent data to text, use the following method:

```
private CharSequence getMessageText(Intent intent) {
    Bundle remoteInput = RemoteInput.getResultsFromIntent(intent);
    if (remoteInput != null) {
        return
        remoteInput.getCharSequence
        (MessagingService.EXTRA_VOICE_REPLY);
    }
    return null;
}
```

The complete `MessageReplyReceiver` class is as follows:

```
public class MessageReplyReceiver extends BroadcastReceiver {

    private static final String TAG =
    MessageReplyReceiver.class.getSimpleName();


    @Override
    public void onReceive(Context context, Intent intent) {
        if (MessagingService.REPLY_ACTION.equals(intent.getAction())) {
            int conversationId = intent.getIntExtra("reply", -1);
            CharSequence reply = getMessageText(intent);
            if (conversationId != -1) {
                Log.d(TAG, "Got reply (" + reply + ") for
                ConversationId " + conversationId);
            }
            // Tell the Service to send another message.
            Intent serviceIntent = new Intent(context,
            MessagingService.class);
            serviceIntent.setAction
            (MessagingService.SEND_MESSAGE_ACTION);
            context.startService(serviceIntent);
```

```
        }
    }

    private CharSequence getMessageText(Intent intent) {
        Bundle remoteInput = RemoteInput.getResultsFromIntent(intent);
        if (remoteInput != null) {
            return remoteInput.getCharSequence
            (MessagingService.EXTRA_VOICE_REPLY);
        }
        return null;
    }
}
```

Afterward, register `broadcast receiver` in the manifest as follows:

```
<receiver android:name=".MessageReplyReceiver">
    <intent-filter>
        <action
        android:name="com.packt.smartchat.ACTION_MESSAGE_REPLY"/>
    </intent-filter>
</receiver>
```

Now, we are completely ready to receive data from the Wear notification component to the local application.

The voice input screen from the Wear notification component will look as follows:

Use the **Draw emoji** on this screen and Android will predict what you have drawn:

IMF can be used to reply by typing the input:

Summary

In this chapter, you have learned how to use Firebase functions to send push notifications and to use the notification component from the Wear support library. Notifications are integral components in smart devices; they play a crucial role by reminding users. We have understood the `NotificationCompat.Builder` class and the `WearableExtender` class. We have also explored the input method framework and the easiest way to reply with multiple reply mechanisms, such as emojis, voice support, and so on.

10
Just a Face for Your Time - WatchFace and Services

The face, also known as the dial, is the part of the clock that displays the time with fixed numbers with moving hands. The appearance of a clock face can be designed with various artistic approaches and creativity. Designing a conventional watch face is a beautiful art; a watch face artist will know what it takes to carve and engineer a watch face for traditional wearable watches. In Android Wear, the process is very similar, except you, being the watch face maker, will not have any tools in your hands, but will instead need to know which service you need to extend and what piece of code will help you customize the look and feel of the watch face. The watch face will show the time and date. Here, in Android Wear, a watch face can be analog or it can be digital.

Android Wear watch faces are services that are packaged inside a wearable app. When users select one of the available watch faces, the wearable device shows the watch face and invokes its service callback methods. Custom watch faces use a dynamic, digital canvas that can incorporate hues, activities, and relevant data. When we install a wearable watch face application in Android Wear, we can switch between different watch faces through the watch face picker. Users can install various watch faces on their watch using the companion application from the Google Play Store on their phones. You will learn the following topics in this chapter:

- The `CanvasWatchFaceService` class and registering your watch face
- The `CanvasWatchFaceService.Engine` and Callback methods
- Writing watch faces and handling gestures and tap events
- Understanding watch face elements and initializing them
- Common issues

The CanvasWatchFaceService class and registering your watch face

Watch faces are services with drawing and visual rendering ability; all watch faces will extend the `CanvasWatchFaceService` class. The `CanvasWatchFaceService` class extracts its functionalities from the `WallpaperSevice` and `WallpaperService.Engine` classes. The `Engine` class, with its callback methods, helps the watch face with its lifecycle. If you have to make a watch face for an Android Wear, you should use the `CanvasWatchfaceService` class instead of plain old vanilla `WallpaperService`. A watch face service, like a wallpaper service, must implement only the `onCreateEngine()` method. Watch face engines need to implement the method `onTimeTick()` to refresh the time and refresh the view and `onAmbientModeChanged(boolean)` to switch between different version of watch faces, such as the grey mode and colorful watch face. Watch face engines in a like manner implement `onInterruptionFilterChanged(int)` to update the view dependent upon how much information the user has inquired. For the updates that occur in the ambient mode, `wake_lock` will be held, so the device doesn't go to rest until the watch face finishes the drawing process. Registering watch faces in the application works closely to registering wallpapers, with a couple of additional steps. However, watch faces require the `wake_lock` permission, which is demonstrated as follows:

```
<uses-permission android:name="android.permission.WAKE_LOCK" />
```

Later, your watch face service declaration needs preview metadata:

```
<meta-data
    android:name="com.google.android.wearable.watchface.preview"
    android:resource="@drawable/preview_face" />
<meta-data
    android:name=
    "com.google.android.wearable.watchface.preview_circular"
    android:resource="@drawable/preview_face_circular" />
```

Lastly, we need to add a special intent filter with the goal that watch.

```
<intent-filter>
    <action android:name="android.service.wallpaper.WallpaperService"
    />
    <category
        android:name=
        "com.google.android.wearable.watchface.category.WATCH_FACE" />
</intent-filter>
```

The CanvasWatchFaceService.Engine class

The `CanvasWatchFaceService.Engine` class extends the `WatchFaceService.Engine` class. Here, actual implementation of a watch face that draws on a canvas can be accomplished. We ought to implement `onCreateEngine()` to reestablish your concrete engine implementation. `CanvasWatchFaceService.Engine` has one public constructor with a couple of procedures to enable us to implement the watch face. How about we examine a couple of methods that we will implement in the later bit of this chapter:

- `void invalidate ()`: Plans a call to `onDraw(Canvas, Rect)` to draw the following frame. This must be approached on the main thread.

- `void onDestroy ()`: In this callback, we can release the hardware and other resources that we would be using to complete the watch face.

- `void onDraw(Canvas canvas, Rect bounds)`: Draws the watch face, all the visual components, and clock revive rationale, and other clock arrangements are accomplished in this method.

- `void onSurfaceChanged()`: This method takes four params, `void onSurfaceChanged (SurfaceHolder holder, int organise, int width, int stature)`. The `SurfaceHolder` parameter enables you to control the surface size and different arrangements.

- `void postInvalidate()`: Posts a message to schedule a call to `onDraw(Canvas, Rect)` to draw the following frame. Furthermore, this method is thread-safe. We can call this method from any thread.

These methods play a noteworthy part in planning your watch face. Let's begin making a watch face. In the following exercise, we will figure out how to make a digital watch face.

Writing your own watch face

Android Studio is the primary tool that we should use to write Wear apps for numerous reasons; since we have already configured our development environment for Wear 2.0 development, it shouldn't be a challenge. Let's fire up Android Studio and create a Wear project.

In Activity chooser, select **Add No Activity**. Since a watch face is a service, we don't need activity:

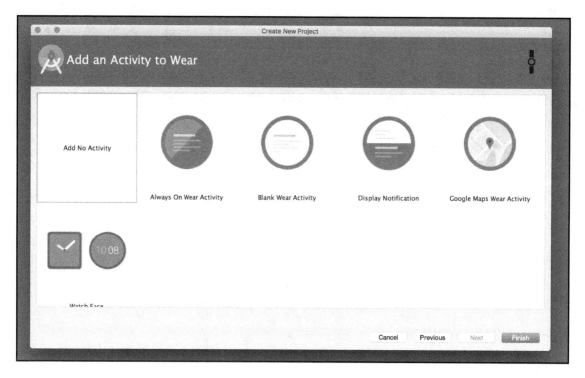

We have read in the previous section that we will be extending the class to CanvasWatchFaceService, where watch face is being drawn on a canvas, and another class is CanvasWatchFaceService.Engine, where we will work on the actual implementation of the watch face and more important methods that we have discussed. It will help us to achieve the necessary setup for the watch face.

Now, let's create a class file called PacktWatchFace in the package.

The `PacktWatchFace` class will extend to the `CanvasWatchFaceService` class:

After creating the class file, extend it to the `CanvasWatchFaceService` class; this is the service class that we will register in the manifest. Inside this class, we need to create one subclass for the Wear design implementation. After that, inside the same class, we need to override an `onCreateEngine()` method.

The following code is the entry point setup for the Wear watch face design:

```
public class PacktWatchFace extends CanvasWatchFaceService{


@Override
public Engine onCreateEngine() {
    return new Engine();
}

private class Engine extends CanvasWatchFaceService.Engine{


    }
}
```

The `PacktWatchFace` class implements only one method `onCreateEngine()` that returns the actual implementation of `CanvasWatchFaceService.Engine`. Now, it is time to register the `watchFace` service class in the manifest file.

Within in the application scope of manifest register, add the `PacktWatchFace` class:

```
<service
    android:name=".PacktWatchFace"
    android:label="PacktWatchface Wear"
    android:permission="android.permission.BIND_WALLPAPER" >
    <meta-data
        android:name="android.service.wallpaper"
        android:resource="@xml/watch_face" />
    <meta-data
        android:name="com.google.android.wearable.watchface.preview"
        android:resource="@mipmap/ic_launcher" />
    <meta-data
        android:name=
        "com.google.android.wearable.watchface.preview_circular"
        android:resource="@mipmap/ic_launcher" />

    <intent-filter>
        <action android:name=
        "android.service.wallpaper.WallpaperService" />

        <category android:name=
        "com.google.android.wearable.watchface.category.WATCH_FACE" />
    </intent-filter>
</service>
```

Create a file under the `xml` folder in the `res` directory and call it `watch_face.xml`. Inside, add the `wallpaper` XML tag, as follows:

```
<?xml version="1.0" encoding="UTF-8"?>
<wallpaper xmlns:android="http://schemas.android.com/apk/res/android"/>
```

Watch face service declaration needs preview metadata, as shown earlier. The same metadata is used in the preview of the watch face picker in wearables. These elements would specify the metadata of watch face service. The watch face will use the preview image and other information that we provide in this metadata tag.

Add the following permission to your manifest:

```
<uses-permission
android:name="com.google.android.permission.PROVIDE_BACKGROUND" />
<uses-permission android:name="android.permission.WAKE_LOCK" />
```

Let's set up the essential instances for graphical elements and chronology instance, globally:

```
//Essential instances
private static Paint textPaint, boldTextPaint, backGround, whiteBackground,
darkText;
private static Calendar calendar;
private static final long INTERACTIVE_UPDATE_RATE_MS =
TimeUnit.SECONDS.toMillis(1);
private static final int MSG_UPDATE_TIME = 0;
```

In the `onDraw` method, we can draw what we need to display on the watch face. The drawn visual is going to be static; we cannot make it dynamic just by drawing on the canvas. The implementation of the chronological time display plays an important role.

What information the watch face needs to show and other cosmetics are decided by the watch face designer. Now, let's initialize all the instances in the `onCreate` method:

```
backGround = new Paint() {{ setARGB(255, 120, 190, 0); }};
textPaint = createPaint(false, 40);
boldTextPaint = createPaint(true, 40);
whiteBackground = createPaint(false, 0);
darkText = new Paint() {{ setARGB(255, 50, 50, 50); setTextSize(18); }};
```

Next, we will write a separate method, which is `createPaint()`, for returning the values of all the calls:

```
private Paint createPaint(final boolean bold, final int fontSize) {
    final Typeface typeface = (bold) ? Typeface.DEFAULT_BOLD :
    Typeface.DEFAULT;

    return new Paint()
    {{
        setARGB(255, 255, 255, 255);
        setTextSize(fontSize);
        setTypeface(typeface);
        setAntiAlias(true);
    }};
}
```

Handling tap events and gestures

On the watch face, users can interact, but the `CanvasWatchService.Engine` class provides only a single interaction method, which is a single tap. If we want to have other interactions, we need to override the `onTapCommand` method. We need to request `tapevents` in the `onCreate` method by changing the style of the Wear application:

```
setWatchFaceStyle(new WatchFaceStyle.Builder(PacktWatchFace.this)
    .setAcceptsTapEvents(true)
    .build());
```

Thereafter, we can override the `onTapCommand()` method to handle the tap events and we can override the function to provide features and services when a user taps on the application.

The following shows toast message when a user clicks on the watch face:

```
@Override
public void onTapCommand(
        @TapType int tapType, int x, int y, long eventTime) {
    switch (tapType) {
        case WatchFaceService.TAP_TYPE_TAP:
            // Handle the tap
            Toast.makeText(PacktWatchFace.this, "Tapped",
            Toast.LENGTH_SHORT).show();
            break;

        // There are other cases, not mentioned here. <a
        href="https://developer.android.com/training/wearables/watch-
        faces/interacting.html">Read Android guide</a>
        default:
            super.onTapCommand(tapType, x, y, eventTime);
            break;
    }
}
```

This way, we can customize the tap functionality. The default function signature gives two coordinates, *x* and *y*; by using these coordinates, we can determine where a user has clicked, which helps watch face designers to customize gestures and tap events accordingly.

Supporting different form factors

Android Wear devices comes in square and rectangle designs. It's a watch face developer's responsibility to make the watch face looks same in both the form factors. Most of the UI arrangements that are designed for rectangular displays will fail on circular displays, and vice versa. To resolve this issue, the WallpaperService Engine has a facility called the onApplyWindowInsets function. The onApplyWindowInsets method helps to check whether the device is round or not; by determining this, we can draw either a round or a square watch face:

```
@Override
public void onApplyWindowInsets(WindowInsets insets) {
    super.onApplyWindowInsets(insets);
    boolean isRound = insets.isRound();

    if(isRound) {
        // Render the Face in round mode.
    } else {
        // Render the Face in square (or rectangular) mode.
    }
}
```

Now, let's write a complete method that draws the watch face with timely updates:

```
@Override
    public void onDraw(Canvas canvas, Rect bounds) {
        calendar = Calendar.getInstance();

        canvas.drawRect(0, 0, bounds.width(), bounds.height(),
        whiteBackground); // Entire background Canvas
        canvas.drawRect(0, 60, bounds.width(), 240, backGround);

        canvas.drawText(new SimpleDateFormat("cccc")
        .format(calendar.getTime()), 130, 120, textPaint);

        // String time = String.format
        ("%02d:%02d", mTime.hour, mTime.minute);
        String time = new SimpleDateFormat
        ("hh:mm a").format(calendar.getTime());
        canvas.drawText(time, 130, 170, boldTextPaint);

        String date = new SimpleDateFormat
        ("MMMM dd, yyyy").format(calendar.getTime());
        canvas.drawText(date, 150, 200, darkText);
    }
```

The `onVisibilityChanged` method helps in registering and unregistering the receiver that tells the time to watch face:

```
@Override
public void onVisibilityChanged(boolean visible) {
    super.onVisibilityChanged(visible);

    if (visible) {
        registerReceiver();
        // Update time zone in case it changed while we weren't
        visible.
        calendar = Calendar.getInstance();
    } else {
        unregisterReceiver();
    }

    // Whether the timer should be running depends on whether we're
    visible (as well as
    // whether we're in ambient mode), so we may need to start or stop
    the timer.
    updateTimer();
}

private void registerReceiver() {
    if (mRegisteredTimeZoneReceiver) {
        return;
    }
    mRegisteredTimeZoneReceiver = true;
    IntentFilter filter = new
    IntentFilter(Intent.ACTION_TIMEZONE_CHANGED);
    PacktWatchFace.this.registerReceiver(mTimeZoneReceiver, filter);
}

private void unregisterReceiver() {
    if (!mRegisteredTimeZoneReceiver) {
        return;
    }
    mRegisteredTimeZoneReceiver = false;
    PacktWatchFace.this.unregisterReceiver(mTimeZoneReceiver);
}

/**
 * Starts the {@link #mUpdateTimeHandler} timer if it should be running and
isn't currently
 * or stops it if it shouldn't be running but currently is.
 */
private void updateTimer() {
    mUpdateTimeHandler.removeMessages(MSG_UPDATE_TIME);
```

```
    if (shouldTimerBeRunning()) {
        mUpdateTimeHandler.sendEmptyMessage(MSG_UPDATE_TIME);
    }
}
```

To make sure the timer is running only when the watch face is visible, we will set the following configuration:

```
private boolean shouldTimerBeRunning() {
    return isVisible() && !isInAmbientMode();
}
```

To update the time periodically in the watch face, do the following:

```
private void handleUpdateTimeMessage() {
    invalidate();
    if (shouldTimerBeRunning()) {
        long timeMs = System.currentTimeMillis();
        long delayMs = INTERACTIVE_UPDATE_RATE_MS
                - (timeMs % INTERACTIVE_UPDATE_RATE_MS);
        mUpdateTimeHandler.sendEmptyMessageDelayed(MSG_UPDATE_TIME,
        delayMs);
    }
}
```

Now, let's finalize the code with the WeakReference class implementation. Weak reference objects will allow referents to be finalized, and can be accessed later. Weak reference will make all of the previous weakly reachable objects to be finalized. Finally, it will en queue those recently cleared weak references that registered with the reference queues:

```
private static class EngineHandler extends Handler {
    private final WeakReference<Engine> mWeakReference;

    public EngineHandler(PacktWatchFace.Engine reference) {
        mWeakReference = new WeakReference<>(reference);
    }

    @Override
    public void handleMessage(Message msg) {
        PacktWatchFace.Engine engine = mWeakReference.get();
        if (engine != null) {
            switch (msg.what) {
                case MSG_UPDATE_TIME:
                    engine.handleUpdateTimeMessage();
                    break;
            }
        }
```

```
        }
    }
```

To add a drawable, we can make use of the `BitmapFactory` class:

```
bitmapObj = BitmapFactory.decodeResource(getResources(),
R.mipmap.ic_launcher);

// And set the bitmap object in onDraw methods canvas
canvas.drawBitmap(bob, 0, 40, null);
```

Now that the complete logic definition is complete, let's see the complete finalized class for watch face:

```
/**
 * Created by ashok.kumar on 27/04/17.
 */

public class PacktWatchFace extends CanvasWatchFaceService{
    //Essential instances
    private static Paint textPaint, boldTextPaint, backGround,
    whiteBackground, darkText;
    private static Calendar calendar;
    private static final long INTERACTIVE_UPDATE_RATE_MS =
    TimeUnit.SECONDS.toMillis(1);
    private static final int MSG_UPDATE_TIME = 0;

    @Override
    public Engine onCreateEngine() {
        return new Engine();
    }

    private class Engine extends CanvasWatchFaceService.Engine {

        final Handler mUpdateTimeHandler = new EngineHandler(this);

        final BroadcastReceiver mTimeZoneReceiver = new
        BroadcastReceiver() {
            @Override
            public void onReceive(Context context, Intent intent) {
                calendar = Calendar.getInstance();
            }
        };
        boolean mRegisteredTimeZoneReceiver = false;

        boolean mLowBitAmbient;

        @Override
```

```
public void onCreate(SurfaceHolder holder) {
    super.onCreate(holder);

    setWatchFaceStyle(new WatchFaceStyle.Builder
    (PacktWatchFace.this)
            .setCardPeekMode(WatchFaceStyle.PEEK_MODE_SHORT)
            .setBackgroundVisibility
            (WatchFaceStyle.BACKGROUND_VISIBILITY_INTERRUPTIVE)
            .setShowSystemUiTime(false)
            .build());

    backGround = new Paint() {{ setARGB(255, 120, 190, 0); }};
    textPaint = createPaint(false, 40);
    boldTextPaint = createPaint(true, 40);
    whiteBackground = createPaint(false, 0);
    darkText = new Paint()
    {{ setARGB(255, 50, 50, 50); setTextSize(18); }};

    setWatchFaceStyle(new WatchFaceStyle.Builder
    (PacktWatchFace.this)
            .setAcceptsTapEvents(true)
            .build());

    calendar = Calendar.getInstance();
}

private Paint createPaint
(final boolean bold, final int fontSize) {
    final Typeface typeface =
    (bold) ? Typeface.DEFAULT_BOLD : Typeface.DEFAULT;

    return new Paint()
    {{
        setARGB(255, 255, 255, 255);
        setTextSize(fontSize);
        setTypeface(typeface);
        setAntiAlias(true);
    }};
}

@Override
public void onTapCommand(
        @TapType int tapType, int x, int y, long eventTime) {
    switch (tapType) {
        case WatchFaceService.TAP_TYPE_TAP:
            // Handle the tap
            Toast.makeText(PacktWatchFace.this,
            "Tapped", Toast.LENGTH_SHORT).show();
```

```
            break;

        default:
            super.onTapCommand(tapType, x, y, eventTime);
            break;
    }
}

@Override
public void onDestroy() {
    mUpdateTimeHandler.removeMessages(MSG_UPDATE_TIME);
    super.onDestroy();
}

@Override
public void onPropertiesChanged(Bundle properties) {
    super.onPropertiesChanged(properties);
    mLowBitAmbient = properties.getBoolean
    (PROPERTY_LOW_BIT_AMBIENT, false);
}

@Override
public void onTimeTick() {
    super.onTimeTick();
    invalidate();
}

@Override
public void onAmbientModeChanged(boolean inAmbientMode) {
    super.onAmbientModeChanged(inAmbientMode);
    if (inAmbientMode) {
        if (mLowBitAmbient) {
        }
        invalidate();
    }

    updateTimer();
}

@Override
public void onDraw(Canvas canvas, Rect bounds) {
    calendar = Calendar.getInstance();

    canvas.drawRect(0, 0, bounds.width(), bounds.height(),
    whiteBackground); // Entire background Canvas
    canvas.drawRect(0, 60, bounds.width(), 240, backGround);

    canvas.drawText(new SimpleDateFormat("cccc")
```

```
            .format(calendar.getTime()), 130, 120, textPaint);

        // String time = String.format("%02d:%02d", mTime.hour,
        mTime.minute);
        String time = new SimpleDateFormat
        ("hh:mm a").format(calendar.getTime());
        canvas.drawText(time, 130, 170, boldTextPaint);

        String date = new SimpleDateFormat
        ("MMMM dd, yyyy").format(calendar.getTime());
        canvas.drawText(date, 150, 200, darkText);
    }

    @Override
    public void onVisibilityChanged(boolean visible) {
        super.onVisibilityChanged(visible);

        if (visible) {
            registerReceiver();
            calendar = Calendar.getInstance();
        } else {
            unregisterReceiver();
        }

        updateTimer();
    }

    private void registerReceiver() {
        if (mRegisteredTimeZoneReceiver) {
            return;
        }
        mRegisteredTimeZoneReceiver = true;
        IntentFilter filter = new IntentFilter
        (Intent.ACTION_TIMEZONE_CHANGED);
        PacktWatchFace.this.registerReceiver
        (mTimeZoneReceiver, filter);
    }

    private void unregisterReceiver() {
        if (!mRegisteredTimeZoneReceiver) {
            return;
        }
        mRegisteredTimeZoneReceiver = false;
        PacktWatchFace.this.unregisterReceiver(mTimeZoneReceiver);
    }

    private void updateTimer() {
        mUpdateTimeHandler.removeMessages(MSG_UPDATE_TIME);
```

```java
                if (shouldTimerBeRunning()) {
                    mUpdateTimeHandler.sendEmptyMessage(MSG_UPDATE_TIME);
                }
            }

            private boolean shouldTimerBeRunning() {
                return isVisible() && !isInAmbientMode();
            }

            private void handleUpdateTimeMessage() {
                invalidate();
                if (shouldTimerBeRunning()) {
                    long timeMs = System.currentTimeMillis();
                    long delayMs = INTERACTIVE_UPDATE_RATE_MS
                            - (timeMs % INTERACTIVE_UPDATE_RATE_MS);
                    mUpdateTimeHandler
                    .sendEmptyMessageDelayed(MSG_UPDATE_TIME, delayMs);
                }
            }
        }

        private static class EngineHandler extends Handler {
            private final WeakReference<Engine> mWeakReference;

            public EngineHandler(PacktWatchFace.Engine reference) {
                mWeakReference = new WeakReference<>(reference);
            }

            @Override
            public void handleMessage(Message msg) {
                PacktWatchFace.Engine engine = mWeakReference.get();
                if (engine != null) {
                    switch (msg.what) {
                        case MSG_UPDATE_TIME:
                            engine.handleUpdateTimeMessage();
                            break;
                    }
                }
            }
        }
    }
}
```

The final, compiled watch face will be available in your Wear device's watch face picker:

Congratulations on building your first watch face.

Understanding watch face elements and initializing them

Depending on what kind of watch face we are developing, we need to plan out certain elements for the watch face. We have seen what it takes to build a digital watch face, but to build an analog watch face, we need to understand a few watch face elements that will help in constructing watch face.

Generally, analog watch face is the combination of three essential components, as follows:

- HOUR_STROKE
- MINUTE_STROKE
- SECOND_TICK_STROKE

Now, to construct an analog watch face, we need these three components and the rest of the things are going to be almost similar as constructing a digital watch face. Here, we need to make a little more effort in animating the strokes.

First, we need to design Strokes, as shown in the following code:

```
mHourPaint = new Paint();
mHourPaint.setColor(mWatchHandColor);
mHourPaint.setStrokeWidth(HOUR_STROKE_WIDTH);
mHourPaint.setAntiAlias(true);
```

```
mHourPaint.setStrokeCap(Paint.Cap.ROUND);
mHourPaint.setShadowLayer(SHADOW_RADIUS, 0, 0, mWatchHandShadowColor);

mMinutePaint = new Paint();
mMinutePaint.setColor(mWatchHandColor);
mMinutePaint.setStrokeWidth(MINUTE_STROKE_WIDTH);
mMinutePaint.setAntiAlias(true);
mMinutePaint.setStrokeCap(Paint.Cap.ROUND);
mMinutePaint.setShadowLayer(SHADOW_RADIUS, 0, 0,
mWatchHandShadowColor);

mSecondPaint = new Paint();
mSecondPaint.setColor(mWatchHandHighlightColor);
mSecondPaint.setStrokeWidth(SECOND_TICK_STROKE_WIDTH);
mSecondPaint.setAntiAlias(true);
mSecondPaint.setStrokeCap(Paint.Cap.ROUND);
mSecondPaint.setShadowLayer(SHADOW_RADIUS, 0, 0,
mWatchHandShadowColor);

mTickAndCirclePaint = new Paint();
mTickAndCirclePaint.setColor(mWatchHandColor);
mTickAndCirclePaint.setStrokeWidth(SECOND_TICK_STROKE_WIDTH);
mTickAndCirclePaint.setAntiAlias(true);
mTickAndCirclePaint.setStyle(Paint.Style.STROKE);
mTickAndCirclePaint.setShadowLayer(SHADOW_RADIUS, 0, 0,
mWatchHandShadowColor);
```

Now, with the previously designed strokes, we can design and customize the watch face the way we want, and can add different backgrounds on the canvas, along with other cosmetic elements that will make your analog watch special.

For a digital watch face, you need a reference for the text and other graphical elements you will be using in the watch face.

Common issues

Wear watch face applications are different than Wear apps. The most common issue that watch face apps encounter is different form factors, such as square and round dial chin. To resolve this issue, programmers have to detect the Wear form factor before the watch face executes. As we have already discussed, the onApplyWindowInsets() method of CanvasWatchFaceService.Engine helps in finding the shape of the Wear.

Watch face apps are always running; essentially, watch face service extends to wallpaper service. When we have a lot of services that fetch data from the network API calls, the battery might drain quickly. Such services might include:

- Different form factors
- Battery efficacy
- UI accommodation
- Too much of animations
- Assets we use to build the Wear watch face
- Watch face depending on hardware

UI accommodation is another challenge for watch face makers; while we enable `setHotwordIndicator()` in the watch face style, the Android system should be able to post and overlay notification cards on top of the watch face we build. The analog watches we build must take care of this scenario, since analog watches are little reluctant to resize and, in the continuous animation of strokes, it will not coordinate with system notifications. Too much of animations in watch face is not a good idea. Having many animations results in CPU and GPU performance issues. The things to consider when we have animations in watch face are as follows:

- Reducing the frame rate of animations
- Letting the CPU sleep between animations
- Reducing the size of bitmaps assets used
- Disabling anti-aliasing when drawing scaled bitmaps
- Moving expensive operations outside the drawing method

When your watch face depends on a hardware to show data, you should make sure you are periodically accessing the hardware and releasing it. For instance, when we are using the GPS to show the current location and the watch face is continuously listening to the GPS, we will not just drain the battery, the garbage collector will also throw an error.

Interactive watch faces

The trend changes every time when a Wear 2.0 update arrives which gives new interactive watch faces, which also can have unique interaction and style expression. All the watch face developers for Wear might have to start thinking of interactive watch faces.

What is exactly an interactive watch face?

The idea is to have the user like and love watch face by giving them delightful and useful information on a timely basis, which changes the user experience about the watch face.

Google addresses the following methodologies to achieve interactive watch faces:

- Creative vision
- Different form factors
- Display modes
- System UI elements
- Data integrated watch faces

Android Wear offers a digital canvas to express time in a very efficient way. Android Wear offers to integrate a data on watch faces for higher level of personalization as well. Watch faces and designs need to be glanceable and should convey prioritized information to the watch face user.

We know that the different form factors of Android Wear is an implementation challenge for the watch face developers. The watch face should be identical across different form factors with regards to its design language, which is a common set of colors, line width, shading, and other design elements.

In Wear 2.0, there are two display modes:

- Active mode
- Always active mode

Active mode is when a user moves their wrist or touches the display to glance the time. Wear will light up the display and make the watch face active. In this mode, the watch face can use colorful animations and fluid design language to express the time and other information.

Always active mode helps to save battery power and the display capabilities are limited to black, white, and gray when the Wear device enters to the always active mode. We need to carefully design what to display in the always active mode that looks similar to the design of the watch face, but with less color and animations.

System UI elements indicate the status of the Wear devices; for instance, battery level and other system UI elements. The watch face should allow these indicators to be displayed in some specific location of Wear device.

Data integrated watch faces help watch face users to check out the chosen information at a glance, for example, step counts, weather reports, and so on can be displayed on the watch face.

Summary

In this chapter, we have explored the fundamental understanding of designing watch faces and we have built a digital watch face. We have understood how the `CanvasWatchFaceService` class helps in building watch faces and we have also seen the following watch face-related concepts:

- The `CanvasWatchFaceService` class
- The `canvasWatchFaceService.Engine` method
- Registering watch face in a Wear module manifest
- Handling tap gesture
- Different form factors
- Adding bitmap images to the watch face
- Watch face elements
- Common issues
- Interactive watch faces

Making watch faces is an excellent artistic engineering, including what data we should express in the watch face and how time and date data is being displayed. The `ComplicationsAPI` is new in Wear 2.0. Let's discuss that in the next chapter along with a few advanced concepts.

11
More About Wear 2.0

Android Wear 2.0 is a prominent update with plenty of new features bundled in, including Google Assistant, standalone applications, new watch faces, and support for third-party complications. In previous chapters, we explored how to write different kinds of Wear applications. Wear 2.0 offers more with the current market research and Google is working with partner companies to build a powerful ecosystem for Wear.

In this chapter, let's understand how we can take our existing skills forward with the following concepts:

- Standalone applications
- Curved layouts and more UI components
- Complications API
- Different navigations and actions
- Wrist gestures
- Input method framework
- Distributing Wear apps to the Play store

Standalone applications

In Wear 2.0, standalone applications are powerful feature of the wear ecosystem. How cool it will be using wear apps without your phone nearby! There are various scenarios in which Wear devices used to be phone dependent, for example, to receive new e-mail notifications, Wear needed to be connected to the phone for Internet services. Now, wear devices can independently connect to Wi-Fi and can sync all apps for new updates. The user can now complete more tasks with wear apps without a phone paired to it.

Identifying an app as a standalone

The idea of a standalone application is a great feature of the wear platform. Wear 2.0 differentiates the standalone app through a metadata element in the Android manifest file. Inside the application tag `<application>`, the `</application>` metadata element is placed with `com.google.android.wearable.standalone` with the value true or false. The new metadata element indicates whether the wear app is a standalone app and doesn't require the phone to be paired to operate. When the metadata element is set to true, the app a can also be available for wear devices working with iPhone. Generally, the watch app can be categorized as follows:

- Completely independent of the phone app
- Semi-independent (without phone pairing, wear app will have limited functionality)
- Dependent on the phone app

To make a Wear app completely independent or semi-independent, set the value of the metadata to true, as follows:

```
<application>
...
  <meta-data
    android:name="com.google.android.wearable.standalone"
    android:value="true" />
...
</application>
```

Now, any platform, such as iPhone or Android phones without the Play store on them, can also use wear apps, downloading them directly from the Play store present in wear. By setting the value of the metadata to false, we tell Android Wear that this Wear app is dependent on a phone with the Play store app.

 Note: Regardless of the possibility that the value is false, the watch application can be installed before the corresponding phone application is installed. In this way, if a watch application identifies that a companion phone does not have a necessary phone application, the watch application ought to incite the user to install the phone application.

Standalone apps storage

You can use standard Android storage APIs to store data locally. For instance, you can use the SharedPreference API, SQLite, or internal storage. We have, up until now, explored how to integrate ORM libraries, such as Realm, into wear applications not simply to store away data, but to likewise share code between a wear application and a phone application. On the other hand, code that is specific to a shape component and form factor can be in a different module.

Detecting wear app on another device

Android wear apps and Android mobile apps can recognize support apps using the Capability API. Phone and wear apps can advertise to paired devices statically and dynamically. At the point when an application is on the node in a user's wear network, the **Capability API** enables another application to identify the correspond installed applications.

Advertise capability

To launch an activity on a handheld device from a wearable device, use the `MessageAPI` class to send the request. Multiple wearables can be associated with the handheld Android device; the wearable application needs to determine that an associated node is fit to launch the activity from a handheld device application. To advertise the capability of the handheld application, perform the following steps:

1. Create an XML configuration file in the `res/values/` directory of your project and name it `wear.xml`
2. Add a resource named `android_wear_capabilities` to `wear.xml`
3. Define the capabilities that the device provides

 Note: Capabilities are custom strings that you characterize and should be unique within your application.

The following example shows how to add a capability named `voice_transcription` to `wear.xml`:

```
<resources>
    <string-array name="android_wear_capabilities">
        <item>voice_transcription</item>
    </string-array>
</resources>
```

Retrieving the nodes with the required capability

Initially, we can detect the capable nodes by calling the `CapabilityAPI.getCapability()` method. The following examples show how to manually retrieve the results of reachable nodes with the `voice_transcription` capability:

```
private static final String
        VOICE_TRANSCRIPTION_CAPABILITY_NAME = "voice_transcription";

private GoogleApiClient mGoogleApiClient;

...

private void setupVoiceTranscription() {
    CapabilityApi.GetCapabilityResult result =
            Wearable.CapabilityApi.getCapability(
                    mGoogleApiClient,
                    VOICE_TRANSCRIPTION_CAPABILITY_NAME,
                    CapabilityApi.FILTER_REACHABLE).await();

    updateTranscriptionCapability(result.getCapability());
}
```

To detect the capable nodes as they connect to a wearable device, register a `CapabilityAPI.capabilityListner()` instance to `googleAPIclient`. The following example shows how to register the listener and retrieve the results of reachable nodes with the `voice_transcription` capability:

```
private void setupVoiceTranscription() {
    ...

    CapabilityApi.CapabilityListener capabilityListener =
            new CapabilityApi.CapabilityListener() {
                @Override
                public void onCapabilityChanged(CapabilityInfo
                capabilityInfo) {
```

```
                    updateTranscriptionCapability(capabilityInfo);
            }
        };

    Wearable.CapabilityApi.addCapabilityListener(
            mGoogleApiClient,
            capabilityListener,
            VOICE_TRANSCRIPTION_CAPABILITY_NAME);
}
```

 Note: On the off chance that you create a service that extends
`WearableListenerService` to identify capability changes, you might
need to override the `onConnectedNodes()` method to listen in to finer-
grained connectivity details, for example, when a wearable device changes
from Wi-Fi to a Bluetooth connection with the handset. For more data on
the most proficient method to listen for important events, read **Data Layer
Events**.

In the wake of recognizing the capable nodes, figure out where to send the message. You
ought to pick a node that is in close proximity to your wearable device to limit message
routing through numerous nodes. A nearby node is characterized as one that is directly
associated with the device. To decide whether a node is nearby, call the `Node.isNearby()`
method.

Detecting and guiding the user to install a phone app

Now, we know how to detect the wear and mobile applications using the Capability API.
It's an opportunity to guide the user to install the corresponding application from the Play
Store.

Use the `CapabilityApi` to check whether your phone application is installed on the paired
phone. For more data, see the Google samples. In the event that your phone application isn't
installed on the phone, use
`PlayStoreAvailability.getPlayStoreAvailabilityOnPhone()` to check what sort
of phone it is.

If `PlayStoreAvailability.PLAY_STORE_ON_PHONE_AVAILABLE` is returned valid, it
implies the phone is an Android phone with the Play store installed. Call
`RemoteIntent.startRemoteActivity()` on the wear device to open the Play Store on
the phone. Use the market URI for your telephone application (which might not be the same
as your phone URI). For instance, use a market URI:

`market://details?id=com.example.android.wearable.wear.finddevices.`

In the event that `PlayStoreAvailability.PLAY_STORE_ON_PHONE_UNAVAILABLE` is returned, it implies the phone is likely an iOS phone (with no Play Store accessible). Open the App Store on the iPhone by calling `RemoteIntent.startRemoteActivity()` on the wear device. You can indicate your application's iTunes URL, for instance, `https://itunes.apple.com/us/application/yourappname`. Likewise, observe opening a URL from a watch. On an iPhone, from Android Wear, you can't programmatically determine whether your phone application is installed. As a best practice, give a mechanism to the user (for example, a button) to manually trigger the opening of the App Store.

Using the `remoteIntent` API portrayed earlier, you can determine that any URL can be opened on the phone from a wear device and no phone application is required.

Getting just the important information

In most use cases, when we get data from the Internet, we just get the necessary information. Any more than that and we may encounter pointless idleness, memory use, and battery use.

At the point when a wear device is associated with a Bluetooth LE association, the Wear application may have the entrance to a data transfer capacity of just 4 kilobytes for each second. Depending on wear, the accompanying steps are prescribed:

1. Review your network requests and responses for additional information, that is for a phone application

2. Shrink huge pictures before sending them over a network to a watch

Cloud messaging

For the tasks identified for notifications, applications can specifically use **Firebase Cloud Messaging** (**FCM**) in Wear applications; Google Cloud informing is not supported in wear 2.0.

There are no particular FCM APIs for wear applications; it takes after the comparative configuration for the mobile application notification: FCM functions admirably with wear and in doze mode. FCM is the recommended approach to send and receive notifications for wear devices.

The procedure for receiving notifications from the server is that the application needs to send the device a unique Firebase `registration_id` to the server. The server can then distinguish the `FCM_REST` endpoint and can send the notification. An FCM message is in the JSON format and can incorporate either of the accompanying payloads:

- **Notification payload**: Generic notification data; when a notification reaches Wear, the application can check the notification and users can launch the application that received the notification.
- **Data Payload**: The payload will have custom key-value sets. The payload will be conveyed as data to wear applications.

Wear applications incorporate many concerns when we are developing applications specific to wear devices, acquiring high-bandwidth networks and reducing the picture quality particular to wear benchmarks. What's more, UI outlines and keeping up background services, and so on. Having this at the top of the priority list when we create applications will make them stand out in the crowd.

Complications API

Complications are surely just the same old thing for watches. The internet says the first pocket watch with complications was revealed in the sixteenth century. Smart watches are the ideal place for all the components that we consider for complications. In Android wear, the watch face shows more data than just the time and date, such as a step counter, climate forecast, and so on. How these complications have functioned so far has had a major constraint. until now, each custom watch face application needed to execute its own rationale to get information to show. For instance, if two watch faces had a comparative component to get step counts and show relevant information, then it would be an exercise in futility. Android Wear 2.0 intends to take care of this issue with the new Complications API.

In the event of complications, a watch face communication data provider assumes the principle part. It incorporates logic to get the information. The watch face won't have immediate access to the data provider; it will get callbacks when there is other data with the selected complications. On the other hand, data providers won't know how the data will be utilized; that is up to the watch face.

The accompanying depiction discusses how watch faces get complications data from the data provider:

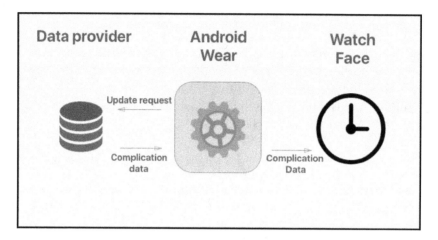

Complication data provider

The new complications API has immense potential; it has access to battery level, climate, step counts, and so on. The complication data provider, which is a service, extends ComplicationProviderService. This base class has a set of callbacks, keeping in mind the end goal, which is to know when a provider is chosen as a data source for the current watchface:

- (onComplicationActivated): This callback method is called when complication is activated.
- (onComplicationDeactivated): This callback method is called when complication is deactivated.
- (onComplicationUpdate): This callback is called when the complication has updated information for the particular complication id.

The ComplicationProviderService class is an abstract class, which extends out to a service. A provider service must implement onComplicationUpdate (int, int and ComplicationManager) to respond to requests for updates from the complication system. The manifest declaration of this service must incorporate an intent filter for ACTION_COMPLICATION_UPDATE_REQUEST. Metadata to determine the supported types, refresh period, and config action, if required, ought to also be incorporated: (METADATA_KEY_SUPPORTED_TYPES, METADATA_KEY_UPDATE_PERIOD_SECONDS, and METADATA_KEY_PROVIDER_CONFIG_ACTION).

The manifest entry for the service ought to likewise incorporate an android: Icon attribute. The icon given there ought to be a single-color white icon that represents the provider. This icon will appear in the provider chooser interface, and may likewise be incorporated into `ComplicationProviderInfo`, given to watch faces to show in their configuration activities.

The accompanying code demonstrates a builder pattern of `ComplicationsData` to a short text type to populate `ComplicationData` with the date of the next event and an optional icon:

```
ComplicationData.Builder(ComplicationData.TYPE_SHORT_TEXT)
    .setShortText(ComplicationText.plainText(formatShortDate(date)))
    .setIcon(Icon.createWithResource(context, R.drawable.ic_event))
    .setTapAction(createContactEventsActivityIntent())
    .build();
```

Adding complications to a watch face

Android Wear 2.0 conveys numerous new components to your smartwatch. Yet, one of the more discernible is the expansion of adaptable *complications* on the watch face. Complications are a two-part system; watch face engineers can plan their watch faces to have open slots for complications, and application developers can surface parts of their applications to incorporate them as complications. The Wear watch face app can receive complication data and enable users to select the data providers. Android wear provides a user interface for data source. We can add complications, or data from applications, to some watch faces. Your wear 2.0 has complications that demonstrate your battery life and the date, and that's just the beginning. You can also include complications from some third-party applications.

Receiving data and rendering complications

To begin receiving complication data, a watch face calls `setActiveComplications` in the `WatchFaceService.Engine` class with a list of watch face complication IDs. A watch face creates these IDs to remarkably identify slots on the watch face where complications can appear, and passes them to the `createProviderChooserIntent` method to enable the user to choose which complication ought to go in which slot. Complication data is conveyed by means of the `onComplicationDataUpdate` (of `WatchFaceService.Engine`) callback.

Allowing users to choose data providers

Android Wear gives a UI (through an activity) that enables users to pick providers for a specific complication. Watch faces can call the `createProviderChooserIntent` method to obtain an intent that can be used to demonstrate the chooser interface. This intent must be used with `startActivityForResult`. At the point when a watch face calls `createProviderChooserIntent`, the watch face supplies a watch face complication ID and a list of supported types.

User interactions with complications

Providers can specify an action that occurs if the user taps on a complication, so it should be possible for most complications to be tap-able. This action will be specified as `PendingIntent`, included in the `ComplicationData` object. The watch face is responsible for detecting taps on complications and should fire the pending intent when a tap occurs.

Permissions for complication data

The Wear app must have the following permission to receive the complications data and open the provider chooser:

```
com.google.android.wearable.permission.RECEIVE_COMPLICATION_DATA
```

Opening the provider chooser

A watch face that has not been granted the preceding permission will be unable to start the provider chooser. To make it easier to request the permission and start the chooser, the `ComplicationHelperActivity` class is available in the Wearable Support Library. This class should be used instead of `ProviderChooserIntent` to start the chooser in almost all cases. To use `ComplicationHelperActivity`, add it to the watch face in the manifest file:

```
<activity
android:name="android.support.wearable.complications.ComplicationHelperActi
vity"/>
```

To start the provider chooser, call the
`ComplicationHelperActivity.createProviderChooserHelperIntent` method to
obtain an intent. The new intent can be used with either `startActivity` or
`startActivityForResult` to launch the chooser:

```
startActivityForResult(
   ComplicationHelperActivity.createProviderChooserHelperIntent(
      getActivity(),
      watchFace,
      complicationId,
      ComplicationData.TYPE_LARGE_IMAGE),
   PROVIDER_CHOOSER_REQUEST_CODE);
```

When the helper activity is started, the helper activity checks whether the permission was
granted. If the permission was not granted, the helper activity makes a runtime permission
request. If the permission request is accepted (or is unneeded), the provider chooser is
shown.

For watch face, there are many scenarios to be considered. Check all of them before you
implement the complications in your watch face. How are you receiving complication data?
Is it from the provider, remote server, or the rest service? Are the provider and watch face
from the same app? You should also check for a lack of appropriate permissions and so on.

Understanding different navigation for wear

Android wear is evolving in every way. In wear 1.0, switching between screens used to be
tedious and confusing to wear users. Now, Google has introduced material design and
interactive drawers, such as:

- **Navigation drawer**: The navigation drawer is a similar component to the mobile
  app navigation drawer. It will allow the user to switch between views. Users can
  reach to the navigation drawer on a Wear device by going to the top of the
  content area and scrolling down from the flings. We can allow the drawer to be
  opened anywhere within the scrolling parent's content by setting the
  `setShouldOnlyOpenWhenAtTop()` method to false and we can restrict it by
  setting it to true.
- **Single page navigation drawer**: A wear app can present the views to users on a
  single page and multiple pages. The new navigation drawer component allows
  content to be on a single page by setting `app:navigation_style to
  single_page`.

- **Action drawer**: There are common actions that every category of apps does. Action drawer provides an access to all such actions in a wear application. Usually, action drawer sits in the bottom area of the wear app and it will help to offer context-specific user actions, similar to the action bar on a phone application. It's the developer's choice to have the action drawer positioned at the bottom or top, and the action drawer can be triggered when a user is scrolling content.

The following image is a quick look of the navigation drawer for wear 2.0:

The following example illustrates the action reply performed in the messenger app using action:

Implementation

To have the newly introduced components in your app, declare your user interface with a `WearableDrawerLayout` object as the root view of your layout. Within `WearableDrawerLayout`, add one more view that implements `NestedScrollingChild` to contain the main content, as well as additional views to contain content for the drawer.

The following XML code illustrates the way in which we can bring life to `WearableDrawerLayout`:

```xml
<android.support.wearable.view.drawer.WearableDrawerLayout
    android:id="@+id/drawer_layout"
    xmlns:android="http://schemas.android.com/apk/res/android"
    xmlns:tools="http://schemas.android.com/tools"
    android:layout_width="match_parent"
    android:layout_height="match_parent"
    tools:deviceIds="wear">

    <android.support.v4.widget.NestedScrollView
        android:id="@+id/content"
        android:layout_width="match_parent"
        android:layout_height="match_parent">

    <LinearLayout
            android:id="@+id/linear_layout"
            android:layout_width="match_parent"
            android:layout_height="match_parent"
            android:orientation="vertical" />

    </android.support.v4.widget.NestedScrollView>

    <android.support.wearable.view.drawer.WearableNavigationDrawer
        android:id="@+id/top_navigation_drawer"
        android:layout_width="match_parent"
        android:layout_height="match_parent"/>

    <android.support.wearable.view.drawer.WearableActionDrawer
        android:id="@+id/bottom_action_drawer"
        android:layout_width="match_parent"
        android:layout_height="match_parent"
        app:action_menu="@menu/action_drawer_menu"/>

</android.support.wearable.view.drawer.WearableDrawerLayout>
```

Single-page navigation drawer

A single-page navigation drawer is faster and more streamlined to different views in the Wear app. To create a single page navigation drawer, apply the attribute `navigation_style="single_page"` to the drawer. For example:

```
<android.support.wearable.view.drawer.WearableNavigationDrawer
        android:id="@+id/top_drawer"
        android:layout_width="match_parent"
        android:layout_height="match_parent"
        android:background="@android:color/holo_red_light"
        app:navigation_style="single_page"/>
```

Now, the next major thing is to populate the data on drawer layout. We can do it in the XML layout with the help of the `app:using_menu` attribute in the drawer layout and by loading the XML file from the menu directory.

Using `WearableDrawerView`, we can design our own custom drawer layout. The following code illustrates the custom drawer layout:

```
<android.support.wearable.view.drawer.WearableDrawerView
        android:layout_width="match_parent"
        android:layout_height="match_parent"
        android:layout_gravity="top"
        android:background="@color/red"
        app:drawer_content="@+id/drawer_content"
        app:peek_view="@+id/peek_view">
        <FrameLayout
            android:id="@id/drawer_content"
            android:layout_width="match_parent"
            android:layout_height="match_parent">
            <!-- Drawer content goes here. -->
        </FrameLayout>
        <LinearLayout
            android:id="@id/peek_view"
            android:layout_width="wrap_content"
            android:layout_height="wrap_content"
            android:layout_gravity="center_horizontal"
            android:paddingTop="8dp"
            android:paddingBottom="8dp"
            android:orientation="horizontal">
            <!-- Peek view content goes here. -->
        <LinearLayout>
    </android.support.wearable.view.drawer.WearableDrawerView>
```

There are major drawer events, such as `onDrawerOpened()`, `onDrawerClosed()`, and `onDraworStateChanged()`. We can create custom events as well; by default, we can use the earlier set of callbacks to listen to drawer activities.

Notifications in wear 2.0

Wear 2.0 updates the visual style and interaction paradigm of notifications. Wear 2.0 introduces expandable notifications, which provide more content area and actions to give the best experience. The visual updates include material design, touch targets of notifications, dark background colors, and a horizontal swipe gesture for notifications.

Inline action

Inline action allows users to perform context-specific actions inside the notification stream card. If the notification is configured for inline action, it will display at the bottom section of notifications. Inline actions are optional; Google recommends it for different use cases wherein a user has to perform a certain action after checking the notification, for example, text message reply and stopping fitness activity. A notification can have only one inline action and, to enable it, we need to set `setHintDisplayActionInline()` to true.

To add inline actions to notifications, perform the following steps:

1. Create an instance of `RemoteInput.Builder` as follows:

```
String[] choices =
context.getResources().getStringArray(R.array.notification_reply_ch
oices);     choices = WearUtil.addEmojisToCannedResponse(choices);
RemoteInput remoteInput = new RemoteInput.
Builder(Intent.EXTRA_TEXT)
.setLabel(context.getString
    (R.string.notification_prompt_reply))
    .setChoices(choices)
    .build();
```

2. Using the `addRemoteInput()` method, we can attach the `RemoteInput` object:

```
NotificationCompat.Action.Builder actionBuilder = new
NotificationCompat.Action.Builder(
    R.drawable.ic_full_reply, R.string.notification_reply,
    replyPendingIntent);
actionBuilder.addRemoteInput(remoteInput);
actionBuilder.setAllowGeneratedReplies(true);
```

3. Finally, add a hint to display action inline, and use add action method to the action to the notification:

```
// Android Wear 2.0 requires a hint to display the reply action
inline.
    Action.WearableExtender actionExtender =
        new Action.WearableExtender()
            .setHintLaunchesActivity(true)
            .setHintDisplayActionInline(true);
    wearableExtender.addAction
    (actionBuilder.extend(actionExtender).build());
```

Expanded notifications

Wear 2.0 introduces expandable notifications, which have the ability to show substantial content and actions for each notification. Expanded notifications follow the material design standards, and when we attach additional content pages to the notification, they are available within the expanded notifications and the user will have an in-app experience while checking the actions and content in the notifications.

Best practices for expanded notifications

When to use the expanded notifications:

1. The notifications from the paired phone should use expanded notifications.
2. We should not use expanded notifications when the app notification is running locally and just launches the app by clicking on it.

Bridging mode for notifications

Bridging mode alludes to the system that wear and the companion application share the notification among themselves. The standalone application and companion application can get copied notifications. Android wear incorporates components to deal with the issue of copy notifications.

Developers can alter the behavior of notifications as follows:

- Specifying a bridging configuration in the manifest file
- Specifying a bridging configuration at runtime
- Setting a dismissal ID so notification dismissals are synced across devices

Bridging configuration in manifest file:

```
<application>
...
  <meta-data
    android:name="com.google.android.wearable.notificationBridgeMode"
    android:value="NO_BRIDGING" />
...
</application>
```

Bridging configuration at runtime (uses the `BridgingManager` class):

```
BridgingManager.fromContext(context).setConfig(
  new BridgingConfig.Builder(context, false)
    .build());
```

Using dismissal ID to sync notification dismissals:

```
NotificationCompat.WearableExtender wearableExtender =
  new NotificationCompat.WearableExtender().setDismissalId("abc123");
Notification notification = new NotificationCompat.Builder(context)
// ... set other fields ...
  .extend(wearableExtender)
  .build();
```

Notification is an important component to draw the attention of users to use your app in wear devices. Android Wear 2.0 offers more, and will continue to offer more, smart replies in notifications, messaging style, and more.

Wear 2.0 input method framework

We have seen the wear input mechanism in the apps we built in previous chapters. Wear 2.0 supports input methods beyond voice by extending the Android **Input Method Framework** (**IMF**) to Android Wear. IMF allows for virtual, onscreen keyboards and other input methods to be used for text entry. The IMF APIs used for wear devices are the same as other form factors, though the usage is slightly different due to the limited screen size. Wear 2.0 comes with the system default **Input Method Editor** (**IME**) and opens up the IMF APIs for third-party developers to create custom input methods for Wear.

Invoking input method for wear

To invoke the IMF for wear, your API level should be 23 or higher on the platform. In Android Wear apps that contain an EditText fields: Touching a text field places the cursor in the field and automatically displays the IMF on focus.

Wrist gestures

Wear 2.0 supports wrist gestures. When you cannot use the touch screen of your wear device, you can utilize the wrist gestures for quick, one-handed interaction, for example, when a user is jogging and he wants to perform a certain context-specific operation using a wrist gesture. There are few gestures that are not available for apps, such as push wrist down, raise wrist up, and shaking wrist. Every wrist gesture is mapped to an integer constant from the key event class:

Gesture	KeyEvent	Description
Flick wrist out	KEYCODE_NAVIGATE_NEXT	This key code goes to the next item.
Flick wrist in	KEYCODE_NAVIGATE_PREVIOUS	This key code goes to the previous item.

Best practices for using gestures in apps

Following are the best practices for using gestures in apps:

- Review the `KeyEvent` and `KeyEvent.Callback` pages for the delivery of key events to your view
- Have a touch parallel for a gesture
- Provide visual feedback
- Don't reinterpret repeated flick gestures into your own new gesture. It may conflict with the system's *Shaking the wrist* gesture.
- Use `requestFocus()` and `clearFocus()` carefully

Authentication protocols

With standalone watches in place, wear apps can now run entirely on a watch without a companion app. This new capability also means that Android Wear standalone apps will need to manage authentication on their own when the apps need to access data from the cloud. Android Wear supports several authentication methods to enable standalone wear apps to obtain user authentication credentials. Now, wear supports the following:

- Google sign-in
- OAuth 2.0 support
- Pass tokens via data layer
- Custom code authentication

All these protocols follow the same standard as mobile app programming; no big changes in integrating Google sign-in in wear or other protocols, but these protocols help in authorizing users.

App distribution

We now know how to develop applications for wear 2.0 and, in our past experience, we might have published an Android app to the Play store. What does it take to publish a standalone wear application, or a general wear application, to the Play store through the Google developer console?

Wear 2.0 bundles the Play Store app; users can search for wear-specific apps and can install them directly on Wear devices when they are connected to the Internet. Generally, wear 2.0 apps in the Play store need a minimum and target an API level of 25 or higher in the manifest file.

Publish your first wear app

To make your app appear in the on-watch Play Store, generate a signed wear apk. Publishing the app will be similar to publishing a mobile app if it's a wear standalone app. If it is not a standalone and you need to upload a multiple apk, then follow `https://develo per.android.com/google/play/publishing/multiple-apks.html`.

Let's publish the Wear-Note app in the Play store. It is Google's dedicated dashboard for developers that lets you manage your apps in the Play store. Google has a one-time $25 registration fee, which you need to pay before you can upload an app. The reason behind the fee is to keep out fake, duplicate accounts and, hence, keep out unnecessary and poor apps flooding the Play Store.

Following steps illustrates the story of how we can publish our wear application to playstore:

    1. Visit `https://play.google.com/apps/publish/`:

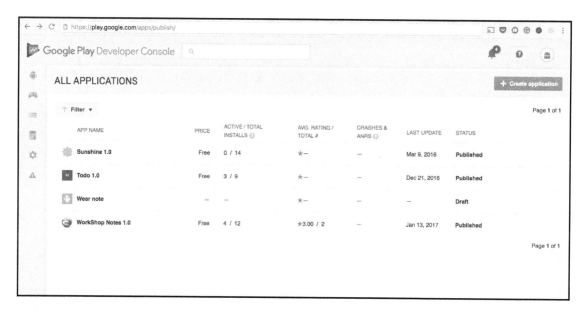

    2. Click on **Create application** and give your app a title.
    3. Thereafter, you will be shown a form to fill in the description and other details, which include screenshots of the app and icons and promo graphics.
    4. In store listing, fill in all the correct information about the app.
    5. Now, upload the signed wear apk.
    6. Answer the questionnaire for content ratings, get a rating, and apply the rating to your app.
    7. In pricing and distribution, you need to have a merchant account to distribute your app in the pricing model. Now, wear note app is a free the Wear app and it lets you select free.

8. Select all the countries on the list and choose wear device apk:

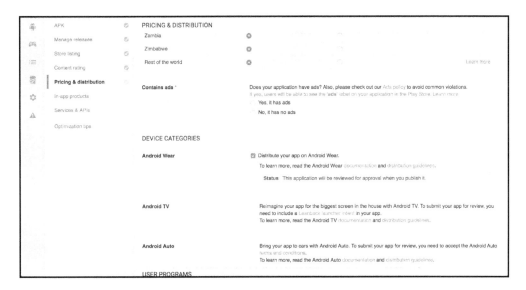

9. Google will review the wear binary and approve to distribute it in the Wear Play store when it is ready.
10. Congratulations! Now, your app is ready to be published:

Summary

In this chapter, we have understood standalone applications and the Complications API. We have seen how to detect companion apps using Capability API, and we have a clear idea of standalone applications and publishing a wear app too.

This chapter examined how we can reinforce the comprehension of wear 2.0 and its components, along with an exhaustive understanding of standalone apps, curved layouts and more UI components, and building Wear applications with navigation drawers and action drawers. It also offered a brief understanding of wrist gestures and using them in wear applications, using the input method framework, and distributing the wear application to the Google Play Store.

Index